Charles Alexander Cameron

The Stock-Feeder's Manual

The Chemistry of Food in Relation to the Breeding and Feeding of Live Stock

Charles Alexander Cameron

The Stock-Feeder's Manual
The Chemistry of Food in Relation to the Breeding and Feeding of Live Stock

ISBN/EAN: 9783337145873

Printed in Europe, USA, Canada, Australia, Japan

Cover: Foto ©berggeist007 / pixelio.de

More available books at **www.hansebooks.com**

THE STOCK-FEEDER'S MANUAL.

THE
CHEMISTRY OF FOOD

IN RELATION TO THE

BREEDING AND FEEDING

OF

LIVE STOCK.

BY

CHARLES A. CAMERON, Ph.D., M.D.,

Licentiate of the King and Queen's College of Physicians in Ireland; Honorary Corresponding Member of the New York State Agricultural Society; Member of the Agricultural Society of Belgium; Professor of Hygiene or Political Medicine in the Royal College of Surgeons; Professor of Chemistry and Natural Philosophy in Steevens' Hospital and Medical College; Lecturer on Chemistry in the Ledwich School of Medicine; Analyst to the City of Dublin; Chemist to the County of Kildare Agricultural Society, the Queen's County Agricultural Society, &c.; Member of the International Jury of the Paris Exhibition, 1867; Editor of the "Agricultural Review;" one of the Editors of the "Irish Farmer's Gazette;" Author of the "Chemistry of Agriculture," "Sugar and the Sugar Duties," &c. &c.

---*---

LONDON AND NEW YORK:
CASSELL, PETTER, AND GALPIN.

1868.

[*All rights reserved.*]

THE FOLLOWING PAGES ARE

Dedicated

TO

THE RIGHT HONORABLE

THE LORD TALBOT DE MALAHIDE, F.R.S.,

President of the Royal Irish Academy,
&c. &c. &c.,

ONE OF THE MOST ENLIGHTENED AND LIBERAL PROMOTERS

OF AGRICULTURAL IMPROVEMENTS.

THE AUTHOR IS UNDER MANY OBLIGATIONS TO HIS LORDSHIP, FOR

WHICH HE CAN MAKE NO RETURN SAVE THIS PUBLIC ACKNOW-

LEDGMENT OF HIS INDEBTEDNESS.

PREFACE.

SOME papers on the Chemistry of Food, read before the Royal Agricultural Society of Ireland and the Athy Farmers' Club, and a few articles on the Management of Live Stock, published in the *Weekly Agricultural Review*, constitute the basis of this Work. It describes the nature of the food used by the domesticated animals, explains the composition of the animal tissues, and treats generally upon the important subject of nutrition. The most recent analyses of all the kinds of food usually consumed by the animals of the farm are fully stated; and the nutritive values of those substances are in most instances given. Some information is afforded relative to the breeds and breeding of live stock; and a division of the Work is wholly devoted to the consideration of the economic production of "meat, milk, and butter."

Within the last twenty years the processes of chemical analysis have been so much improved, that the composition of organic bodies is now determined with great accuracy. The analyses of foods made from twenty to fifty years ago, possess now but little value. In this Work the analyses of vegetables quoted are chiefly those recently performed by the distinguished Scotch chemist, Dr. Thomas Anderson, and by Dr. Voelcker. The Author believes that in no other Work of moderate size are there so many analyses of food substances given, and ventures to hope that the success of this Work may fully justify the belief that a "handy" book containing such information as that above mentioned, is much required by stock feeders.

102, *Lower Baggot Street, Dublin*,
    APRIL, 1868.

TABLE OF CONTENTS.

                                                                                            PAGE

INTRODUCTION: History of Agriculture—Agricultural Statistics—Imports of Live Stock 1

PART I. ON THE GROWTH AND COMPOSITION OF ANIMALS.

SECTION I. ANIMAL AND VEGETABLE LIFE. Functions of Plants. Animal Life.—SECTION II. COMPOSITION OF ORGANIC SUBSTANCES. Elements of Organic Bodies. Proximate Composition of Organic Substances.—SECTION III. USE OF FAT IN THE ANIMAL ECONOMY. Fatty Food necessary in Cold Climates. Fat Equivalents.—SECTION IV. RELATION BETWEEN THE COMPOSITION OF AN ANIMAL AND THAT OF ITS FOOD. Tables of Experimental Results.—SECTION V. RELATION BETWEEN THE QUANTITY OF FOOD CONSUMED BY AN ANIMAL AND THE INCREASE OF ITS WEIGHT, OR OF THE AMOUNT OF ITS WORK. Weights of Foods necessary to sustain a Man's Life for twenty-four hours. Value of Manure 8

PART II. ON THE BREEDING AND BREEDS OF STOCK.

SECTION I. THE BREEDING OF STOCK.—SECTION II. THE BREEDS OF STOCK. The Form of Animals. *Breeds of the Ox.* Shorthorns. Devons. Herefords. Ayrshires. Polled Cattle. Kyloes. Long-horned. Kerrys. Alderneys. *Sheep.* The Leicester. Lincoln. Cotswold. Cheviot. Southdown. Shropshire. Blackfaced. *Breeds of the Pig.* Berkshire. Yorkshire. *Breeds of the Horse.* Clydesdales. Suffolk Punch. Hunters and Racers 47

PART III. ON THE MANAGEMENT OF LIVE STOCK.

SECTION I. THE OX. Breeding Cows. Wintering of Young Stock. Shelter of Stock. Milch Cows. Stall Feeding. Cost of Maintaining Animals. Cooking and Bruising Food. Value for Feeding Purposes of various Foods. Bedding Cattle.—SECTION II. THE SHEEP. Breeding Ewes. Yeaning. Rearing of Lambs. Sheep Feeding. Sheep Dips.—SECTION III. THE PIG. Young Pigs. Store Pigs. Fattening Pigs.—SECTION IV. THE HORSE. Foals. Dietaries for the Horse . . 74

PART IV. MEAT, MILK, AND BUTTER.

SECTION I. MEAT. Quality of Meat. Is very Fat Meat Unwholesome? Diseased Meat.—SECTION II. MILK. Composition of Milk of Different Animals. Yield of Milk. Preserved Milk. —SECTION III. BUTTER. History of Butter. Irish Butter. Composition of Butter. The Butter Manufacture . . . 112

PART V. ON THE COMPOSITION AND VALUE OF VEGETABLE FOODS.

SECTION I. THE MONEY VALUE OF FOOD SUBSTANCES.—SECTION II. PROXIMATE CONSTITUENTS OF VEGETABLES. Starch. Sugar. Inulin. Gum. Pectin. Cellulose. Oils and Fats. Stearin. Margarin. Olein. Palmitin. Albumen. Fibrin. Legumin. — SECTION III. GREEN FOOD. The Grasses. Schrœder Brome. Tussac Grass. The Clovers. Leguminous Plants—Vetch, Sainfoin, &c. The Yellow Lupine. Rib Grass Plantain. Ergot in Grasses. Holcus Saccharatus. Green Rye. Buckwheat. Rape. Mustard. Comfrey. Chicory. Yarrow. Melons and Marrows. Cabbage. Furze.—SECTION IV. STRAW AND HAY. *Straw.* Anderson's, Voelcker's, and Cameron's Analyses of Straws. Feeding Experiments with Straw. Relative Values of Straw and Oil-cake. *Hay.* Composition of the Hay of different Grasses. Over-ripening of Hay. Damaged Hay and Straw.—SECTION IV. ROOTS AND TUBERS. *Turnips.* Swedish. White Globe. Aberdeen Yellow. Purple-top. Norfolk Bell. Greystone. Turnip Tops. Analyses of Turnips. Mangel Wurtzel. Chemistry of the Mangel. Stripping Leaves off the Mangel. Beet-root. Parsnip. Carrot. Kohl-rabi. Analyses of Kohl-rabi. Radish. The Radish as a Field Crop. Composition of Radish. Jerusalem Artichoke: Advantages of Cultivating it. Analysis of Jerusalem Artichoke. Potato: Analyses of six varieties. Feeding Value of Potatoes.—SECTION VI. SEEDS. *Wheat.* Analyses of Wheat, Flour, Bran, and Husks. Over-ripening of Grain. Wheat a Costly Food. Analyses of Barley, Oat Grain, Indian Corn, Rye, Rice, Rice-dust, and Buckwheat. Malted Corn. Voelcker's Analyses of Malt and Barley. Experiments of Thompson, Lawes, &c., with Malt. Malt Combings. *Leguminous Seeds.* Beans. Composition of Common Beans, Foreign Beans, Peas. Lentils and Winter Tares. *Oil Seeds.* Rape Seeds. Experiments with Rapeseed. Flax Bolls. Composition of Linseed, Rape-seed, Hemp-seed, and Cotton-seed. Fenugreek Seed.—SECTION VII. OIL-CAKES AND OTHER ARTIFICIAL FOODS. Composition of Linseed, Rape-seed, Cotton-seed, and Poppy-seed Cake. Linseed-cake. Adulteration of Linseed-cake. Rape-cake. Feeding Experiments with Rape-cake. Adulterations of Rape-cake. Cotton-seed Cake. Analyses of Decorticated Cotton-seed Cake. Palm-nut Meal: its Composition and Nutritive Properties. Locust, or Carob Bean: its Composition. Dates. Brewers' Dregs and Distillery Wash. Molasses and Treacle.—SECTION VIII. CONDIMENTAL FOOD. Lawes' Experiments with Thorley's Food. Analyses of Condimental Food. Formula for a Tonic Food.—SECTION IX. TABLES OF THE ANALYSES OF THE ASHES OF PLANTS 147

APPENDIX. AGRICULTURAL STATISTICS. Numbers of Live Stock in the United Kingdom. Value of the Agriculture Products of Great Britain . . . 254

THE CHEMISTRY OF FOOD.

INTRODUCTION.

WHEN Virgil composed his immortal "Bucolics," and Varro indited his profound Essays on Agriculture, the inhabitants of the British Islands were almost completely ignorant of the art of cultivating the soil. The rude spoils torn from the carcasses of savage animals protected the bodies of their hardly less savage victors; and the produce of the chase served almost exclusively to nourish the hardy frames of the ancient Celtic hunters. In early ages wild beasts abounded in the numerous and extensive forests of Britain and Ireland; but men were few, for the conditions under which the maintenance of a dense population is possible did not then exist. As civilisation progressed, men rapidly multiplied, and the demand for food increased. The pursuit of game became merely the pastime of the rich; and tame sheep and oxen furnished meat to the lowly as well as to the great. Nor were the fruits of the earth neglected; for during the latter days of the dominion of the Romans, England raised large quantities of corn. Gradually the food of the people, which at first was almost purely animal, became chiefly vegetable. The shepherds, who had supplanted the hunters, became less numerous than the tillers of land; and the era of tillage husbandry began.

At present the great mass of the rural population of these countries subsist almost exclusively upon vegetable aliment—a diet which poverty, and not inclination, prescribes for them. Were the flesh of animals the staple food of the British peasantry, their numbers would not be nearly so large as they now are, for a given area of land is capable of sustaining a far larger number of vegetarians than of meat eaters. The Chinese are by no means averse to animal food, but they are so numerous, that they are in general obliged to content themselves on a purely vegetable diet.

In the manufacturing districts of Great Britain, there are several millions of people whose condition in relation to food is somewhat different from that of the small farmer and agricultural laborer. The artizans employed in our great industries are comparatively well paid for their toil; and the results of their labor place within their reach a fair share of animal food. This section of the population is rapidly increasing, and consequently is daily augmenting the demand for meat. The rural population is certainly not increasing; rather the reverse. Less manual labor is now expended in the operations of agriculture, and even horses are retiring before the advance of the steam plough. The only great purely vegetable-feeding class is diminishing, and the upper, the middle, and the artizan classes—the beef and mutton eating sections of society—are rapidly increasing. It is clear, then, that we are threatened with a revival of the pastoral age, and that in one way, at least, we are returning to the condition of our ancestors, whose staple food consisted of beef, mutton, and pork.

And here two questions arise. How long shall we be able to supply the increasing demand for meat? how long shall we be able to compete with the foreign feeders? These are momentous queries for the British farmer, and I trust they may be solved in a satisfactory manner. At any time during the present century the foreign or colonial grower of wheat could have undersold the British producer of that article, were the latter not protected by a tariff; but cattle could not, as a general rule, be imported into Great Britain at a cheaper rate than they could be produced at home. Were there no corn imported, it is certain that the price of bread would be greater than it is now, even if the grain harvests had been better than they have been for some years past. A bad cereal harvest in England raises the price of flour, but only to a small and strictly limited extent, because, practically, there is no limit to the amount of bread-stuffs procurable from abroad. When, on the contrary, the turnip crop fails, or that excessive drought greatly curtails the yield of grass, the price of meat and butter increases greatly, and is but slightly modified by the importation of foreign stock.

Hitherto the difficulty of transit has been so great that we have only derived supplies of live stock from countries situated at a short distance, such as Holstein and Holland. Vast herds of cattle are fed with but little expense in America, and myriads of sheep are maintained cheaply in Australia; but the immense distances which intervene between our country and those remote and sparsely populated regions have, hitherto, prevented the superabundant supply of animal food produced therein from being available to the teeming population of the British Isles. Should, however, any cheap mode of conveying live stock, or even their flesh, from those and similarly circumstanced countries be devised, it might render the production of meat in Britain a far less profitable occupation than it is now. That we are increasing the area from whence we draw our supplies of live stock is evident from the fact, that within the last two years enormous numbers of horned stock have been imported from Spain. In that extensive country there are noble breeds of the ox; and it would appear that very large numbers of animals could be annually exported, without depriving the inhabitants of a due supply of bovine meat. As Spain is not very distant, it is likely that this traffic will be increased, and that in a short time we shall be as well supplied with Spanish beef as we are now provided with French flour. Meat is at present dear, and is likely to continue so for some time; but still it is evident that, sooner or later, the British feeders will come into keen competition with the foreign producer of meat, and that the price of their commodity will consequently fall. The mere probability of such a state of things, were there no other reason, should induce the feeder to devote increased attention to the improvement of his stock, and to discover more economical methods of feeding them. There is still much to be learned relative to the precise nutritive values of the various feeding stuffs. The proper modes of cooking, or otherwise preparing, food, are still to be satisfactorily determined; and there are many very important questions in relation to the breeding of stock yet unanswered.

It is but fair to admit that the farmer is earnestly endeavouring to improve his art, and that he is willing, nay anxious, to obtain the co-operation of scientific men, in order to increase his knowledge of the theory as well as the practice of his ancient calling. Indeed, he not only admits the utility of science in agriculture, but often places an undue degree of value upon the theories of the chemist, of the botanist, and of the geologist. This is encouraging to the men of science; but, on the other hand, they must admit that by far the greater portion of the sum of human knowledge has been derived from the experience and observation of men utterly unacquainted with science, in the ordinary signification of that term. This portion of our knowledge is also, in its practical application, the most valuable. In the most important branch of industry—agriculture—the labors of the purely scientific man have as yet borne but scant fruit; whilst the unaided efforts of the husbandman have reclaimed from sterility extensive tracts, and caused them to "blossom as the rose." That practical men should have done so much, and scientific men so little, for agriculture, may easily be explained. Countless millions of men, during many thousands of years, have incessantly been occupied in improving the processes of mechanical agriculture, which, as an *art*, has consequently been brought to a high degree of perfection: but scientific agriculture is a creation of almost our own time, and the number of its cultivators is, and always has been, very small; all its theories cannot, therefore, justly claim that degree of confidence which, as a rule, is only reposed in the opinions founded on the experience of practical workers in the field and in the feeding-house. Still, the farmer has derived a great amount of useful information from the chemist and physiologist; and they alone can explain to him the causes of the various phenomena which the different branches of his art present. There was a time when it was the fashion of the man of science to look down with contempt, from the lofty pedestal on which he placed himself, upon the lessons of practical experience read to him by the cultivator of the soil; whilst at the same time the farmer treated as foolish visionaries those who applied the teachings of science to the improvement of their art. But this time has happily passed away. The scientific man no longer despises the knowledge of the mere farmers, but turns to good account the information derivable from their experience; whilst the farmer, on the other side, has ceased to speak in contemptuous terms of mere "book learning." It is to this happy combination of the theorist with the practical man that the recent remarkable advance in agriculture is chiefly due; and to it we may confidently look for improvement in the economic production of meat and butter, and for the enlargement of our knowledge of the relative value of food substances.

STATEMENT OF THE NUMBER OF LIVE STOCK IN GREAT BRITAIN AND IRELAND.

	Enumerated, 1866.			Estimated, 1865.		
	Cattle.	Sheep.	Pigs.	Cattle.	Sheep.	Pigs.
England	3,307,034	15,124,541	2,066,299	3,422,165	18,691,088	2,363,724
Wales ..	541,401	1,668,663	191,604	—	—	—
Islands..	17,700	57,685	22,887	—	—	—
Scotland	937,411	5,255,077	219,716	974,437	5,683,168	146,354
Ireland .	3,493,414	3,688,742	1,299,893	3,493,414	3,688,742	1,299,893
Total	8,316,960	25,794,708	3,800,399	7,890,016	28,062,998	3,809,971

STATEMENT OF THE POPULATION AND NUMBER OF LIVE STOCK IN THE UNITED KINGDOM AND VARIOUS FOREIGN COUNTRIES, ACCORDING TO THE LATEST RETURNS.

Countries.	Date of Returns of Live Stock.	Population according to Latest Returns.	Cows.	Cattle. Other Cattle.	Total.	Sheep.	Pigs.
United Kingdom	1865-66	29,070,932	3,286,308	5,030,652	8,316,960	25,795,708	3,802,399
Russia	1859-63	74,139,394			25,444,000	45,130,800	10,097,000
Denmark Proper	1861	1,662,734	756,834	361,940	1,118,774	1,751,950	300,928
Sleswig	1861	421,486	217,751	172,250	390,001	362,219	87,867
Holstein	1861	561,831	198,310	92,062	290,372	165,344	82,398
Sweden	1860	3,859,728	1,112,944	803,714	1,916,658	1,644,156	457,981
Prussia	1862	18,491,220	3,382,703	2,251,797	5,634,500	17,428,017	2,709,709
Hanover	1861	1,880,070			949,179	2,211,927	554,056
Saxony	1861	2,225,240	411,563	226,897	638,460	371,986	270,462
Wurtemburg	1861	1,720,708	466,758	490,414	957,172	683,842	216,965
Grand Duchy of Baden	1861	1,429,199	348,418	273,068	621,486	177,322	307,198
,, ,, Hesse	1863	853,315	187,442	129,211	316,653	231,787	195,596
,, ,, Nassau	1864	468,311	116,421	84,224	200,645	152,584	65,979
,, ,, Mecklenb. Schwerin	1857	539,258	197,622	69,215	266,837	1,198,450	157,522
,, Oldenburg	1852	279,637			219,843	295,322	87,336
Holland	1864	3,618,459	943,214	390,673	1,333,887	930,136	294,636
Belgium	1856	4,529,461			1,257,649	583,435	458,413
France	1862	37,386,313	5,781,465	8,415,895	14,197,360	33,281,592	5,246,403
Spain	1865	15,658,531			2,904,598	22,054,967	4,264,817
Austria	1863	36,267,648	6,353,086	7,904,030	14,257,116	16,964,236	8,151,608
Bavaria	1863	4,807,440	1,530,626	1,655,356	3,185,882	2,058,638	926,522
United States	1860	31,445,080	8,728,862	8,182,813	16,911,475	23,317,756	32,555,267

INTRODUCTION.

NUMBERS OF THE LIVE STOCK IMPORTED INTO GREAT BRITAIN DURING THE ELEVEN MONTHS ENDED 31ST NOVEMBER, 1867.

Bullocks, bulls, and cows	150,518
Calves	20,720
Sheep and lambs	504,514
Pigs	45,566
	721,318

AMOUNT OF ANIMAL FOOD IMPORTED DURING SAME PERIOD.

Bacon and hams	cwts.	452,132
Salt beef	,,	163,638
Salt pork	,,	123,257
Butter	,,	1,000,095
Lard	,,	213,599
Cheese	,,	798,267
Eggs		373,042,000

I am indebted to Professor Ferguson, Chief of the Veterinary Department of the Irish Privy Council Office, for the following statement:—

RETURN OF HORNED CATTLE EXPORTED FROM THE SEVERAL IRISH PORTS AT WHICH VETERINARY INSPECTORS HAVE BEEN APPOINTED, AND CERTIFIED AS FREE FROM DISEASE, FROM THE 18TH OF NOVEMBER, 1866, TO THE 16TH OF NOVEMBER, 1867 (52 WEEKS).

Fat Stock	187,483
Store Stock	317,331
Breeding and Dairy Stock	36,599
Total	541,413

PART I.

ON THE GROWTH AND COMPOSITION OF ANIMALS.

SECTION I.

ANIMAL AND VEGETABLE LIFE.

Functions of Plants.—It is the primary function of plants to convert the inorganic matter of the soil and air into organised structures of a highly complex nature. The food of plants is purely mineral, and consists chiefly of water, carbonic acid, and ammonia. Water is composed of the elements oxygen and hydrogen; carbonic acid is a compound of oxygen and carbon; and ammonia is formed of hydrogen and nitrogen. These four substances are termed the *organic elements*, because they form by far the larger portion—sometimes the whole—of organic bodies. The combustible portion of plants and animals is composed of the organic elements; the incombustible part is made up of potassium, sodium, and the various other elements enumerated in another page. The organic elements are furnished chiefly by the atmosphere, and the incombustible matters are supplied by the soil.

Water in the state of vapor forms, according to the temperature and other conditions of the atmosphere, from a half per cent. to four and a half per cent. of the weight of that fluid—about 1·25 per cent. being the average; carbonic acid exists in it to the extent of 1-2000th; and ammonia forms a minute portion of it—according to Dr. Angus Smith, one grain weight in 412·42 cubic feet of air (of a town), or 0·000453 per cent. It is remarkable that the most abundant constituents of atmospheric air—oxygen and nitrogen—are not assimilable by plants, although these elements enter largely into the composition of vegetable substances. In the soil, also, the part which ministers to the wants of vegetables is relatively quite insignificant in amount.

Plants are unendowed with organs of locomotion, their food must therefore be within easy reach. Every breeze wafts gaseous nutriment to their expanded leaves, and their rootlets ramify throughout the soil in search of appropriate mineral aliment. But no matter how abundant, or however easy of reach may be the food of plants, the vegetable organism is incapable of partaking of it unless under the influence of light. Exposed to this potent stimulus, the plant collects the gaseous carbonic acid and the vaporous water, solidifies them, decomposes them, and combines their elements into new and organised forms. In effecting these changes—in conferring vitality upon the atoms of lifeless matter—the plant acts merely as the *mechanism*, the light is the *force*. As the work performed by the steam-engine is proportionate to the amount of force developed by the combustion of the fuel beneath its boiler, so is the rapidity of the elaboration of organic substances by plants proportionate to the amount of sunlight to which they are exposed. It is an axiom that matter is indestructible; we may alter its form as often as we please, but we cannot destroy a particle of it. It is the same with *force*: we may convert one kind of it into another —heat into light, or magnetism into electricity—but our power ends there; we can only cause force, or *motion*, to pass from one of its conditions to another, but its *quantity* can never be diminished by the power of man.

The principle of the Conservation of the Forces gives us a clear explanation of the fact that animals can obtain their food only through the medium of the vegetable kingdom. Plants are stationary mechanisms; they have no need to develop motive power, as animals have, in moving themselves from place to place. Their temperature is, we may say, the same as that of the medium in which they exist. Such beings as plants do not, therefore, require the expenditure of force to maintain their vitality; on the contrary, their mechanisms are, for a beneficent purpose, constructed for the *accumulation* of force. The growing plant absorbs, together with carbonic acid, water, and ammonia, a proportionate amount of light, heat, and the various other subtle forces which have their abiding place in the sunbeam—

"That golden chain,
Whose strong embrace holds heaven and earth and main."

Co-incidentally with the conversion of the mineral constituents of the food of plants into organised structures—albumen, fibre, and such like substances—the light, and the heat, and the various other forces likewise suffer a change. Although the precise nature of the new force into which they are converted is still a mystery—one, too, which may never be revealed to us—still we know sufficient of it to satisfy us that it can only exist in connection with organic or organised structures. It is owing to its presence that the elements of these structures (the natural state of which is mineral) are bound together in what may be aptly designated a constrained state; or, as Liebig aptly expresses it, like the matter in a bent spring. So long as the organic structure retains its form, it will be a reservoir of latent force—which will manifest itself in some form during the recoil of the atoms of the matter forming the structure to their original mineral, or statical condition: so the bent spring, when the pressure is removed, returns to its original straight form.

Animal Life.—The chief manifestation of the life of a plant is the accumulation of force; very different are the functions of animal life. It is only by the continuous *expenditure* of force that the vitality of animals is preserved; the heat of a man's body, his power of locomotion, the performance of his daily toil, even his very faculty of thought, are all dependent upon, and to a great extent proportionate to, the amount of organised matter disorganised in his body. It is by the conversion of this organised matter into its original mineral state of water, carbonic acid, and ammonia, that the force originally expended in arranging, through the agency of plants, its atoms, is again restored, chiefly in the form of heat and animal motive power.

Animals, as a class, are completely dependent upon vegetables for their existence. There is every reason to believe that the most lowly organised beings in the scale of animal life, even those of so simple a structure as to have been long regarded as vegetables or as plant-animals, are incapable of organising mineral matter. The so-called vegetative life of animals—for I believe the term to be exceedingly inexact—is applied to their growth, that is, to the increase in their weight. This increase takes place by their power of reorganising, or of assimilating to the nature of their own organisms, certain of the substances elaborated by plants, and destined to become food for animals.

SECTION II.

COMPOSITION OF ORGANIC SUBSTANCES.

Elements of Organic Bodies.—The number of distinct kinds of substances—each distinguishable from all the others by the peculiarity of its properties, taken as a whole—is exceedingly great, yet all these substances are resolvable into a very small number of bodies. As an illustration, I shall take a well-known substance, common green copperas, or, as the chemists term it, protosulphate of iron. By submitting this compound to the process termed chemical analysis, two other kinds of matter may be obtained from it, namely, oxide of iron and oil of vitriol, or sulphuric acid. If we continued this process—if we submitted the acid and the oxide to analysis— we could separate the former into sulphur and oxygen, and the latter into iron and oxygen. Now, by these means we could demonstrate the compound nature of copperas; we could prove that it was *proximately* composed of sulphuric acid and oxide of iron; and, *ultimately*, of iron, sulphur, and oxygen.

Iron, sulphur, and oxygen, are elementary, or simple bodies. They cannot be decomposed; they cannot be analysed. Torture them as we will in our crucibles; expose them as we please to the highest temperature of a wind furnace, or to the more intense heat evolved by a powerful galvanic battery; subject them to the influence of any agent, or force, or process we may choose, and still they will yield nothing but iron, sulphur, and oxygen: hence these undecomposable bodies are regarded as *elements*, or simple substances. So far as our knowledge extends, there are about sixty-six of these undecomposable bodies, of which about one half occurs in but exceedingly minute quantities, and a considerable number of the others exists in comparatively small amounts. As by far the greater proportion of compounds is made up of two or more of about a dozen elementary bodies, it would at first sight appear as if the distinct kinds of compounds which exist, or which may be called into existence by the chemist, must be limited to, at most, a realisable number; but the fact is there is no practical limit to the variety of substances which may be artificially formed. Every difference in the mode of the arrangement of the constituent atoms of a compound, causes its metamorphosis into another kind of substance. To prove that the number of these changes is bounded by no narrow limits, I need but refer to the rules of Permutation, which demonstrate that twelve letters of the alphabet may be arranged in no fewer than 479,000,000 different ways.* The elements are the letters of

---

* If the elements were only capable of combining with each other in simple ratios, the number of their combinations would be as limited as that of the letters of the alphabet; but as one, two, or more atoms of oxygen can combine with one, two, or more atoms of other elements, we can assign no limits to the number of *possible* combinations. There are hundreds o distinct substances formed of but two elements, namely, hydrogen and carbon.

Nature's alphabet, their compounds are the words of the language of Creation. The combinations of sounds and of signs which express the ideas and sensations of man may be limited to millions; but numberless are the hieroglyphs by which the Divine wisdom and beneficence is inscribed on the pages of the magnificent volume of Nature.

Of the sixty-six elementary bodies, not more than a dozen occur commonly in animal and vegetable substances; these are Oxygen, Hydrogen, Nitrogen, Carbon, Sulphur, Phosphorus, Chlorine, Silicium, Potassium, Sodium, Calcium, Magnesium, and Iron. In addition to these, Iodine, and sometimes Bromine, are found in plants which grow in or near the sea; and the former element has also been detected in some of the lower animals, and in land plants. Manganese, Lithium, Cæsium, Rubidium, and a few others of the simple bodies, occasionally occur in plants and animals, but I believe their presence therein is always accidental.

Proximate Composition of Animal Substances.—The differences between vegetable and animal substances are often more apparent than real. Indeed many of the more important of these substances are almost identical in composition. The albumen which coagulates when the juices of vegetables are boiled, is identical with the albumen of the white of eggs; the fibrine of wheat is in no respect chemically different from the fibrine, or clot, of the blood; and, lastly, the legumine, or *vegetable caseine*, of peas is almost undistinguishable from the curd of milk, or *animal caseine*. But not only has chemical research demonstrated the identity of the albumen, fibrine, and caseine of vegetables with three of the more important constituents of animals, it has gone a step further, and proved that they differ from each other in but a few unimportant respects. They are unquestionably convertible into each other* within

---

* In a paper by Professor Sullivan, of Dublin, the conversion of one of these substances into another *outside* the animal mechanism, is almost incontrovertibly proved.

the animal organism; and their functions, as elements of nutrition, are almost, if not quite, identical.

Exclusive of the blood, which contains the elements of every part of the body, the animal organism is composed of three distinct classes of substances—namely, *nitrogenous*, *non-nitrogenous*, and *mineral*. All of these constituents, or substances capable of being converted into them, must exist in the food. Certain articles, for example, milk, contains all of them; but in others, for instance, butter, only one of these substances is found. The nitrogenous part of the body embraces the muscles, or lean flesh, the gelatine of the bones, and the skin and its appendages—such as hair and horns; the non-nitrogenous constituents are its fat and oil; and its mineral matter is found chiefly in the bony framework. These constituents are not, however, isolated: the mineral matter, no doubt, accumulates in certain parts, but in small quantities it is found in every portion of the body; and although the fat forms a distinct tissue, the muscles of the leanest animal are never free from a sensible proportion of it.

Albumen, fibrine, and caseine are the principal nitrogenous constituents of food, and as they are employed in the reparation of the nitrogenous tissues of the animal body, they have been termed *flesh-formers*.

The fat and oil of animals are derived either from vegetable oil and fat, or from some such substance as starch or sugar. The constituents of food which form fat are termed *fat-formers;* and sometimes *heat-givers* or *respiratory elements*, from the notion that their slow combustion in the animal body is the chief cause of its high temperature.

The mineral elements of the body are furnished principally by the varieties of food which contain nitrogen. The whey of milk is rich in them; but they do not exist in pure butter, in starch, or in sugar.

Fat is a much more abundant constituent of the animal body than is generally supposed. That this substance should constitute the greater portion of the weight of an obese pig seems probable enough; but few are aware that even in a lean sheep there is 50 per cent. more fat than lean.

For a very accurate knowledge of the relative proportions of the fatty, nitrogenous, and mineral constituents of the carcasses of animals used as human food, we are indebted to Messrs. Lawes and Gilbert. Before these investigators turned their attention to this subject, it had scarcely attracted the notice of scientific men; but a notion appears to have been current, amongst non-scientific people, at least, that in all, save the fattest animals, the lean flesh greatly preponderated over the fat. That this idea was unsustained by a foundation of fact, has been clearly proved by the results of an investigation* undertaken a few years ago by Messrs. Lawes and Gilbert—an investigation which I cannot avoid characterising as one of the most laborious and apparently trustworthy on record. The mere statement of the results of this inquiry occupies 187 pages of one of the huge volumes of the Transactions of the Royal Society—a fact which best indicates the immensity of the labour which these gentlemen imposed upon themselves, and which, independently of their other and numerous contributions to scientific agriculture, entitles their names to most honourable mention in the annals of science.

I shall now briefly advert to a few of the more important facts established by Lawes and Gilbert. From a large number of oxen, sheep, and pigs, on which feeding experiments were being conducted, ten individuals were selected. These were, a fat calf, a half-fat ox, a moderately fat ox, a fat lamb, a store sheep, a half-fat old sheep, a fat sheep, a very fat sheep, a store pig, and a fat pig. These animals were

---

* *Experimental Inquiry into the Composition of some of the Animals Fed and Slaughtered as Human Food.* By John Bennet Lawes, F.R.S., F.C.S., and Joseph Henry Gilbert, Ph.D., F.C.S. *Philosophical Transactions of the Royal Society.* Part II., 1860.

killed, and the different organs and parts of their bodies were separately weighed and analysed. The results were, that, with the exception of the calf, all the animals contained, respectively, more fat than lean. The fat ox and the fat lamb contained each three times as much fat as lean flesh, and the proportion of the fatty matters to the nitrogenous constituents of the carcass of the very fat sheep was as 4 to 1. In the pig the fat greatly preponderated over the lean; the store pig containing three times as much, and the fat pig five times as much fat as lean.

That part of the animal which is consumed as food by man, is termed the *carcass* by the butcher, and contains by far the greater portion of the fat of the animal. The *offal*, in the language of the butcher, constitutes those parts which are not commonly consumed as human food, at least by the well-to-do classes. In calves, oxen, lambs, and sheep, the offal embraces the skin, the feet, and the head, and all the internal organs, excepting the kidneys and their fatty envelope. The offal of the pig is made up of all the internal organs, excepting the kidneys and kidney fat. It is the relative proportion of fat in the carcasses analysed by Lawes and Gilbert that I have stated; but as the nitrogenous matters occur in greatest quantity in the offal, it is necessary that the relative proportions of the constituents of the body, taken as a whole, should be considered. On an average, then, it will be found that a fat fully-grown animal will contain 49 per cent. of water, 33 per cent. of dry fat, 13 per cent. of dry nitrogenous matter—muscles separated from fat, hide, &c.— and 3 per cent. of mineral matter. In a lean animal the average proportions of the various constituents will be 54 per cent. of water, $25\frac{1}{2}$ per cent. dry fat, 17 per cent. of dry nitrogenous substances, and $3\frac{1}{2}$ per cent. of mineral matter. In the following table these proportions are set forth.

SUMMARY OF THE COMPOSITION OF THE TEN ANIMALS—SHOWING THE PERCENTAGES OF MINERAL MATTER, DRY NITROGENOUS COMPOUNDS, FAT, TOTAL DRY SUBSTANCE, AND WATER.

1st. In Fresh Carcass. 2nd. In Fresh Offal (equal Sum of Parts, excluding Contents of Stomachs and Intestines). 3rd. In Entire Animal (Fasted Live-weight, including therefore the weight of Contents of Stomachs and Intestines).

Description of Animal.	Per cent. in Carcass.					Per cent. in Offal.					Per cent. in Entire Animal.					Contents of viscera.	Water.
	Mineral matter.	Dry nitrogenous compounds.	Fat.	Dry substance.	Water.	Mineral matter.	Dry nitrogenous compounds.	Fat.	Dry substance.	Water.	Mineral matter.	Dry nitrogenous compounds.	Fat.	Dry substance.	Water.		
Fat calf ...	4·48	16·6	16·6	37·7	62·3	3·41	17·1	14·6	35·1	64·9	3·80	15·2	14·8	33·8	3·17	63·8	
Half-fat ox ...	5·56	17·8	22·6	46·0	54·0	4·05	20·6	15·7	40·4	59·6	4·66	16·6	19·1	40·3	8·19	51·5	
Fat ox ...	4·56	15·0	34·8	54·4	45·6	3·40	17·5	26·3	47·2	52·8	3·92	14·5	30·1	48·5	5·98	45·5	
Fat lamb ...	3·63	10·9	36·9	51·4	48·6	2·45	18·9	20·1	41·5	58·5	2·94	12·3	28·5	43·7	8·54	47·8	
Store sheep ...	4·36	14·5	23·8	42·7	57·3	2·19	18·0	16·1	36·3	63·7	3·16	14·8	18·7	36·7	6·00	57·3	
Half-fat old sheep ...	4·13	14·9	31·3	50·3	49·7	2·72	17·7	18·5	38·9	61·1	3·17	14·0	23·5	40·7	9·05	50·2	
Fat sheep ...	3·45	11·5	45·4	60·3	39·7	2·32	16·1	26·4	44·8	55·2	2·81	12·2	35·6	50·6	6·02	43·4	
Extra fat sheep ...	2·77	9·1	55·1	67·0	33·0	3·64	16·8	34·5	54·9	45·1	2·90	10·9	45·8	59·6	5·18	35·2	
Store pig ...	2·57	14·0	28·1	44·7	55·3	3·07	14·0	15·0	32·1	67·9	2·67	13·7	23·3	39·7	5·22	55·1	
Fat pig ...	1·40	10·5	49·5	61·4	38·6	2·97	14·8	22·8	40·6	59·4	1·65	10·9	42·2	54·7	3·97	41·3	
Means of all ...	3·69	13·5	34·4	51·6	48·4	3·02	17·2	21·0	41·2	58·8	3·17	13·5	28·2	44·9	6·13	49·0	
Means of 8 of the half-fat, fat, and very fat animals...	3·75	13·3	36·5	53·6	46·4	3·12	17·4	22·4	42·9	57·1	3·23	13·3	29·9	46·4	6·26	47·3	
Means of 6 of the fat, and very fat animals	3·38	12·5	39·7	55·4	44·6	3·03	16·9	24·1	44·0	56·0	3·00	12·7	32·8	48·5	5·48	46·0	

SECTION III.

USE OF FAT IN THE ANIMAL ECONOMY.

As fat forms so large a portion of the body, it is evident that the part it plays in the animal economy must be a most important one. The general opinion which prevails amongst scientific men as to its physiological functions was originated by the celebrated Liebig. According to his theory, the food of animals includes two distinct kinds of substances—*plastic** and *non-plastic*. The plastic materials are composed of carbon, hydrogen, oxygen, nitrogen, and a little sulphur and phosphorus. Albumen, fibrine, and casein are plastic elements of nutrition; they form the lean flesh, or muscles, the membranes, and cartilages, the gelatine of the bones, the skin, the hair, and, in short, every part of the body which contains nitrogen. The *non-plastic* elements of nutrition include fat, oil, starch, sugar, gum, and certain constituents of fruits, such as pectine.

All non-plastic substances—and of each kind there are numerous varieties—are capable of conversion, in the animal mechanism, into fat and oil. The non-plastic food substances do not contain nitrogen, hence they are commonly termed non-nitrogenous elements. The oily and fatty matters contain a large proportion of carbon, their next most abundant component is hydrogen, and they contain but little oxygen. Unlike the plastic elements, they are—except the fats of the brain and nervous tissue—altogether destitute of sulphur and phosphorus. The starchy, saccharine, and gummy substances are composed of the same elements as the fatty bodies, but they contain a higher proportion of oxygen.

---

* From the Greek *plasso*, "to form." Plastic materials are sometimes termed *formative* elements; both terms imply the belief that they are capable of giving shape, or form, not only to themselves, but also to other kinds of matter not possessed of formative power.

According to Liebig, fat is used in the animal economy as a source of internal heat. We all know that it is a most combustible body, and that during its inflammation the most intense heat is developed. It is less evident, but not less true, that heat is evolved during its slow oxidation, or decay.

The more rapidly a body burns, the greater is the amount of heat evolved by it in a *given time;* but the total amount of heat developed by a specific weight of the body is the same, whether the combustion takes place rapidly or slowly. An experiment performed with phosphorus illustrates the case perfectly. If we burned two pieces of equal weight, the one in oxygen, the other in atmospheric air, we should find that the former would emit a light five times as brilliant as that evolved by the latter, for the simple reason that its combustion would be five times as rapid. The white, vapor-like matter into which phosphorus is converted by its combustion, is termed *phosphoric acid*. It is composed of phosphorus and oxygen. In forming an ounce of this compound, by the direct oxidation, or combustion of phosphorus, the amount of force, either as heat, or as heat and light, evolved is precisely the same, whether the time expended in the process be a minute or a month.* If, in the experiment I have described, we were to substitute two pieces of fat for the fragments of phosphorus, the results would be precisely similar. The fat burned in oxygen gas would emit intense light and heat; but the total amount of these forces evolved would be neither greater nor less than that developed during

---

* The slow conversion of phosphorus into phosphoric acid takes place in the animal organism; its gradual oxidation in the open air gives rise only to an imperfectly oxidised body—*phosphorous acid*. But the latter fact does not invalidate the general proposition, that the heat emitted by a substance undergoing the process of oxidation is proportionate to the amount of oxygen with which it combines, and is not influenced by the length of time occupied by the process, further than this, that if the oxidation be *very* rapidly effected, a portion of the heat will be converted into an *equivalent* amount of light.

the slower and therefore less brilliant combustion of the fat in ordinary atmospheric air. Now, as we can demonstrate that an ounce of fat will emit a certain amount of heat, if burned within a minute of time, and that neither a larger nor a smaller amount will be developed if the combustion of the fat extend over a period of five minutes, I think we may fairly assume that the amount of heat evolved by the complete oxidation of a specific quantity of fat is constant under all conditions, except, as I have already explained, at high temperatures, when a portion of the heat is converted into light.

In the animal organism fat is burned. The process of combustion no doubt is a very slow one, but still the total amount of heat evolved is just the same as if the fat were consumed in a furnace. When the fat constituting a candle is burned, what becomes of it? Its elements, carbon and hydrogen (we may disregard its small amount of oxygen) combine with the oxygen of the air, and form carbonic acid gas and water. What becomes of the fat consumed within the animal body? It also is converted into carbonic acid gas and water. It is not difficult to prove these statements to be facts. A candle will not burn in atmospheric air which has been deprived of its oxygen, because there is no substance present with which the elements of the taper can combine, consequently the process of combustion cannot go on. Now, a man may in one respect be compared with this taper. He is partly made up of fat; that fat is consumed by the oxygen of the air, and the heat developed thereby keeps the body warm. In the process of respiration oxygen is introduced into the lungs, and from thence, by means of the blood vessels, is conveyed throughout every part of the body. In some way, at present not thoroughly understood, the elements of the fat combine with the oxygen, and are converted into carbonic acid gas and water, which are exhaled from the lungs and from the surface of the body.

Fat is a constituent of both animals and plants. The animal derives a portion of its fat directly from the vegetable;

but it possesses the power of forming this substance from other organic bodies, such, for example, as starch. Plants elaborate fat directly from the minerals—carbonic acid gas, and water.

I have already explained that the growth of plants is, *cæteris paribus*, directly proportionate to the amount of sunlight to which they are exposed. Not less certainly is the force which constitutes the sun-beam expended in grouping mineral atoms into organic forms, than is the heat which converts water into steam. But in neither case is the force destroyed. When the vaporous steam is condensed into the liquid water, all the heat is restored, and becomes palpable. By the ultimate decomposition of vegetable substances all the force expended on their production is liberated, and, in some form, becomes manifest.

When the fat formed in the mechanisms of plants is decomposed in the animal organism, two results follow:— The atoms of the fat are re-converted to their original mineral, or statical conditions of carbonic acid gas and water; and the force which maintained them in their organic state is set free as heat, and its equivalent, motive power.

One of the most useful instruments which the ingenuity of man has devised, is the Thermometer. It is so familiarly known that I need not describe it. This instrument does not enable us to estimate the actual quantity of heat contained in a substance, but it indicates the proportion of that subtile element which is *sensible*—that is recognisable by the sense of touch. The dusky Hindu, clad in his single cotton garment, and the Laplander in his suit of fur, are placed under the most opposite conditions in relation to the heat of the sun—the Indian is exposed during the whole year to Sol's most ardent beams, whilst but a scant share of its genial rays goes to warm the body of the Laplander. Now, if we placed the bulb of a thermometer beneath the tongue of a Hindu, we would find the mercury to stand at 98 degrees on Fahrenheit's scale, and if we repeated the experiment on a Laplander, we would obtain an identical result. Numerous experiments of this nature have been made on individuals in most parts of the world, and the results have proved that the temperature of the blood of man is 98 degrees Fahrenheit, whether he be in India or at Nova Zembla, on the *steppes* of Russia, or the elevated *plateaus* of America. This invariability* of the temperature of the bodies of men and of all other warm-blooded animals, appears the more wonderful when it it is considered that the range of the temperature of the medium in which they exist exceeds 200 degrees Fahrenheit. In India, the mercury in the thermometer has been observed to stand at 145 degrees in the direct sunlight, and at 120 degrees in the shade. In high latitudes the temperature is sometimes so low as 100 degrees below zero. A Russian army, in an expedition to China, in 1839, was exposed for several successive days to a temperature of 42 degrees below zero, and suffered severely in consequence.

The facts which I have cited clearly prove that the animal body possesses the power of generating, or, to speak more correctly, liberating heat, either from portions of its own mechanism or from substances placed within that mechanism.

At one time it was the general belief amongst physiologists that one portion of the food consumed by an animal was employed in repairing the waste of its body, and the remaining part was burned as fuel, evolving heat just in the same way as if it had been consumed in a furnace. It was this theory that led to the classification of food into flesh-formers, and heat-givers. It is now doubted if any portion of the food be really burned in this way; and I, for one, think it far more probable that, before its conversion into carbonic acid gas and water (whereby, according to this theory, it develops the heat which keeps the body warm), it first becomes assimilated, that is,

---

* This statement is not absolutely correct, but the range of variation is confined within such narrow limits as to be quite insignificant.

becomes an integral part of the animal body—blood, fat, muscle. Perhaps we would be nearer the truth if we were to assume that heat is evolved during the decomposition of both the nitrogenous and fatty constituents of the body.

The constantly recurring contractions of the muscles must alone be a source of much heat. The development of animal motive power is said to be strictly proportionate to the amount of muscular tissue decomposed. As the nitrogen of the latter is almost completely excreted under the form of urea, the quantity of the latter daily eliminated from the body of an animal is a measure of the decomposed muscular tissue, and consequently of the amount of muscular power generated in the animal organism.* The correspondence between the amount of the motive power of an animal, and the quantity of effete nitrogen excreted from the body, is limited to laboring men and to the lower animals. Strange as it may appear, it is an incontrovertible fact that men whose pursuits require the constant exercise of the intellectual faculties— lawyers, writers, statesmen, students, scientific men, and other brain-workers—excrete more urea than do men engaged in the most physically laborious occupations. An activity of thoughts and ideas involves a corresponding destruction of the tissues, and these require, for their reparation, the consumption of food. Here, then, we have a physical meaning for the common expression—"food for thought."

That the amount of heat developed in the animal organism, is proportionate to the quantity of fatty matters (or of substances capable of forming them) supplied to it in the shape of food, is a proposition which admits of easy demonstration. The natives of warm regions do not require the generation of much heat within their bodies, because the temperature of the medium in which they exist is generally as high as, or higher than, that of their blood. But as they must consume food for

---

* Doubt has recently been thrown on the truth of this belief by Frankland, Fick, and Wislicenus.

the purpose of repairing the waste of their nitrogenous tissues, and as every kind of food contains heat-producing elements, an excess of heat is developed within their bodies, which, if allowed to accumulate, would speedily produce fatal results. The means by which nature removes this superabundant heat are admirably simple, as indeed all its contrivances are. The skin is permeated with millions of pores, and through these openings a large quantity of vapor is given off, and carries with it the surplus heat. The pores are the orifices of minute convoluted tubes which lie beneath the skin, and when straightened measure each about the tenth of an inch, or, according to a writer in the *British and Foreign Medico-Chirurgical Review* (1859, page 349), the one-fifteenth of an inch in length. According to Erasmus Wilson, the number of these tubes which open into every square inch of the surface of the body is 2,800. The total number of square inches on the surface of an average sized man is 2,500, consequently the surface of his body is drained by not less than twenty-eight miles of tubing, furnished with 7,000,000 openings. The cooling of the body, by the evaporation of water from it, admits of explanation by well-known natural laws. Water, in the state of vapor, occupies a space 1,700 fold greater than it does in its liquid condition. It is heat which causes its vaporous form, but it ceases to be heat when it has accomplished this change in the condition of the liquid; for, suffering itself an alteration, it passes into another form of force—mechanical, or motive power. The heat generated within the body is absorbed by the liquid water, the conversion of the latter into vapor follows, and both the heat and the water, in their altered forms, escape through the pores.

Fatty food necessary in cold climates.—As a grave objection against the chemical theory of heat, it has been urged that rice—the pabulum of hundreds of millions of the inhabitants of tropical regions—contains an exceedingly high proportion of heat-giving substances. I have, however, great doubt as to rice ever forming the exclusive food of those people, without their health being impaired in consequence of the deficiency in that substance of the plastic elements of nutrition. Indeed I believe it is a great mistake to assert that the natives of India live almost exclusively on rice. This article, no doubt, forms a large proportion of their food, but it is supplemented with pulse (the produce of leguminous plants), which is rich in flesh-forming materials, also with dried fish, butter, and various kinds of vegetable and animal food rich in nitrogen. The innutritious nature of rice is clearly shown by its chemical composition, and so large a quantity of it must the Hindu consume in order to repair the waste of his body, that his stomach sometimes acquires prodigious dimensions; hence the term "pot-bellied," so often applied to the Indian ryot. I doubt very much, however, if the stomach of the Hindu, large as it is, could accommodate a quantity of rice, the combustion of which would produce a very excessive development of heat. This substance, when cooked, contains a high proportion of water, the evaporation of which carries off a large amount of the heat generated by the combustion of its respiratory constituents. The amount of motive power developed by the Hindu is small as compared with that which the European is capable of exerting; hence he has less necessity for a highly nitrogenous diet. On the whole, then, I am disposed to think that the food of the natives of tropical climates contains sufficient nitrogenous matters to effectually build up and keep in repair their bodies; it also appears clear to me that the amount of heat developed in their bodies is not excessive, and that it is readily disposed of in converting the water, which enters so largely into their diet, into vapor. The proportion of plastic to non-plastic elements in the diet of the Hindu and of the well-fed European, is probably as follows:—

	Nitrogenous.	Non-nitrogenous (calculated as starch.)
Hindu	1 to	9
European	1 to	8

This statement does not quite correspond with Liebig's, who estimates the proportion of nitrogenous to non-nitrogenous substances in rice as 10 to 123, in beef as ten to seventeen, and in veal as ten to one. The results of Lawes and Gilbert's investigations, already alluded to, have, however, dispelled the illusion that the plastic constituents of flesh exceed its non-plastic. In the potato, which at one time constituted more of the food of the Irish peasantry than rice does that of the Hindu, the proportion of plastic to non-plastic materials is as 10 to 110. The results of some analyses of the food grains consumed in the Presidency of Madras, made by Professor Mayer, of the University of Madras, clearly prove that the food of the inhabitants of that part of India is of a far more highly nitrogenous character than is generally supposed. That the Hindu, who subsists exclusively on rice, exhibits all the symptoms of deficient nutrition, is a fact to which numerous competent observers have testified.

A slight consideration of the facts which I have mentioned leads to the conclusion that the food of the inhabitants of very cold regions is required to produce a large amount of heat. Melons, rice, and other watery vegetable productions, however delicious to the palate of the Hindu, would be rejected with disgust by the Esquimaux, whilst the train oil, blubber, and putrid seal's flesh which the children of the icy North consider highly palatable, would excite the loathing of the East Indian. On this subject I may appositely quote the following remarks by Dr. Kane, the Arctic explorer :—"Our journeys have taught us the wisdom of the Esquimaux appetite, and there are few among us who do not relish a slice of raw blubber, or a chunk of frozen walrus beef. The liver of a walrus (awuktanuk), eaten with little slices of his fat—of a verity it is a delicious morsel. Fire would seem to spoil the curt, pithy expression of vitality which belongs to its uncooked juices. Charles Lamb's roast pig was nothing to awuktanuk. I wonder that raw beef is not eaten at home. Deprived of extraneous fibre, it is neither indigestible nor difficult to masticate. With acids

USE OF FAT IN THE ANIMAL ECONOMY. 27

and condiments, it makes a salad which an educated palate cannot help relishing; and as a powerful and condensed heat-making and anti-scorbutic food, it has no rival. I make this last broad assertion after carefully considering its truth. The natives of South Greenland prepare themselves for a long journey, by a course of frozen seal. At Upper Navik they do the same with the narwhal, which is thought more heat-making than the seal; while the bear, to use their own expression, is 'stronger travel than all.' In Smith's Sound, where the use of raw meat seems almost inevitable from the modes of living of the people, walrus holds the first rank. Certainly this pachyderm (Cetacean?) whose finely condensed tissue and delicately permeating fat (oh! call it not blubber) assimilate it to the ox, is beyond all others, and is the best *fuel* a man can swallow." The gastronomic capabilities of the Esquimaux and of other northern races, and their fondness for fatty food, are exhibited in a sufficiently strong light in the following statements:—

Captain Parry weighed and presented to an Esquimaux lad the following articles:—

	lb. oz.
Frozen seahorse flesh	4 4
Wild seahorse flesh	4 4
Bread and bread dust	1 12
Rich gravy soup	1 4
Water	10 0
Strong grog	1 tumbler.
Raw spirits	3 wine glasses.

This large quantity of food, which the lad did not consider excessive, was consumed by him within twenty-four hours. According to Captain Cochrane a reindeer suffices but for one repast for three Yakutis, and five of them will devour at a sitting a calf weighing 200lbs. Mr. Hooper, one of the officers of the *Plover*, in his narrative of their residence on the shores of Arctic America, states that "one of the ladies who visited them was presented, as a jest, with a small tallow candle, called a purser's dip. It was, notwithstanding, a very pleasant joke to the damsel, who deliberately munched it up with evident relish, and finally drew the wick between her set teeth to clean off any remaining morsels of fat."

The partiality for certain kinds of food, and disgust at other varieties, which particular races of men exhibit, is an instinct which they cannot avoid obeying. Instead of exciting our disgust, as it too frequently does, it should exalt our admiration of the infinite wisdom of the Creator, who by simply adapting man's desire for particular kinds of food to the external conditions under which he is placed, enables him to occupy and "subdue the earth" from the Equator to the Poles.

The food of human beings and of the lower animals who inhabit cold countries is nearly exclusively composed of animal substances. The flesh, fat, and oil of animals occupy less space than do the corresponding elements of vegetables; consequently the nutriment they afford is more concentrated, and a larger quantity can be stowed away without inconvenience in the stomach. The heat-forming constituents of these substances constitute not only the chief part of their bulk, but they are also capable of evolving a greater amount of heat than any other of the respiratory elements. One pound of dry fat will develop as much heat as two and a half pounds of dry starch, and the fattest flesh includes four times as much plastic materials as rice. The diet of people all over the world, unless under circumstances which prevent the gratification of the natural appetite, establishes the intimate relation which subsists between cold and food. The appetite of man is at a minimum at the Equator, and at a maximum within the Arctic circle. The statements as to the voracity of Hottentots and Bosjesmans, recorded in the narratives of travellers, do not in the slightest degree affect the general rule that more is eaten in cold climates than in hot regions. These are mere records of gluttony, and it would not be difficult to find parallel cases in our own country. Gluttony is an abnormal appetite, and the greater part of the food devoured under its unnatural, and generally unhealthy stimulus is not applied to the wants of the body.

The bodies of animals are heated masses of matter, and are subject to the ordinary laws of *radiation*. Every substance radiates its heat, and receives in return a portion of that emitted from surrounding bodies. If two bodies of unequal temperature be placed near each other, the warmer of the two will radiate a portion of its heat to the colder, and will receive some of the heat of the latter in return ; but as the warmer body will emit more heat than it will receive, the result will be, that after a time, the length of which will depend on the nature of the bodies, both will acquire the same temperature. In very warm climates the bodies of animals derive from the sun, and from the heated bodies surrounding them, more heat than they give in return; and were it not for their internal cooling apparatus, which I have described, the heat so absorbed would prove fatal. In every climate, on the contrary, where the temperature is lower than 98°, or " blood heat," the bodies of animals lose more heat by radiation than they receive by the same means. The philosophy of the *clothing* of men and the *sheltering* of the lower animals is now evident. It is not only necessary that heat should be developed within the body, but also that its wasteful expenditure should be prevented. The latter is effected by interposing between the warm body and the cold air some substances (such as fur or wool) which do not readily permit the transmission of heat—*non-conductors* as they are termed. The close down of the eider duck is destined to protect its bosom from the chilling influence of the icy waters of the North Polar Sea, and the quadrupeds of the dreary Arctic Circle are sheltered by thick fur coverings from the piercing blasts of its long winter.

Fat Equivalents.—Whilst it is quite certain that neither nerves nor muscles can be elaborated exclusively out of fat, starch, sugar, or any other non-nitrogenous substance, it is almost equally clear that fat may be formed out of nitrogenous tissue. The quantity of fat, however, which is produced in the animal mechanism, from purely nitrogenous food appears to be relatively very small. No animal is capable of subsisting solely on muscle-forming materials, no matter how abundantly supplied. The food of the Carnivora contains a large proportion of fat, and the nutriment of the Herbivora is largely made up of starch and other fat-formers. Dogs, geese, and other animals fed exclusively upon albumen or white of egg rapidly decreased in weight, and after presenting all the symptoms of starvation, died in three or four weeks.* The fat of the bodies of the Carnivora is almost entirely formed—and probably with little if any alteration—from the fatty constituents of their food. Herbivorous animals, on the contrary, derive nearly all their fat from starch, sugar, gum, cellulose, and other non-nitrogenous, but not fatty, materials.

Although starch is convertible into fat, it is not to be understood that a pound weight of one of these bodies is equivalent to an equal quantity of the other. During the conversion of starch into fat, the greater number of its constituent atoms is converted into water and carbonic acid gas. The greater number of the more important metamorphoses of organised matter, which take place in the animal organum, is the result of either oxidation or fermentation : in the conversion of starch or sugar into fat or oil, both of these processes, it is stated, take place; a portion of the hydrogen is converted by oxidation into water, and by fermentation carbonic acid gas is formed, which removes both oxygen and carbon. Perhaps in the formation of fat fermentation is alone employed—a portion of the oxygen being removed as water, and another portion as carbonic acid. The chief difference between the ultimate composition of starch

---

* The results of Savory's experiments on rats appear to prove that animals can live on food destitute of fat, sugar, starch, or any other fat-forming substance. I think, however, that animals could hardly thrive on purely nitrogenous food. The conclusions which certain late writers, who object to Liebig's theory of animal heat, have deduced from Savory's investigations, appear to me to be quite unfounded.

and fat is, that the latter contains a much larger proportion of hydrogen and carbon. The knowledge of the exact quantity of starch required for the formation of a given amount of fat is of importance in enabling us to estimate the relative feeding value of both substances. Certain difficulties stand in the way of our acquiring an accurate knowledge on this point. Not only are there several distinct kinds of fat, but the precise formula, or atomic constitution of each, is as yet veiled in doubt. There are three fats which occur in man and the domesticated animals, and in vegetables. These are stearine, margarine, and oleine. The relative proportions of these vary in each animal : thus, in man and in the goose margarine is the most abundant fat, whilst oleine* exists in the pig in a greater porportion than in man, the sheep, or the ox. The composition of the animal fats does not, however, vary much ; and this fact, together with other considerations, have led chemists to assume that two-and-a-half parts of starch are required for the production of one part of the mixed fats of the different animals. Grape sugar and the pectine bodies — substances which form a large proportion of the food of the Herbivora— contain more oxygen and hydrogen than exist in starch, and, consequently, are not capable of forming so large an amount of fat as an equal weight of starch. We may assume, then, that 2·50 parts of starch, 2·75 parts of sugar, or 3 parts of the pectine bodies, are equivalent to 1 part of fat.

SECTION IV.

RELATION BETWEEN THE COMPOSITION OF AN ANIMAL AND THAT OF ITS FOOD.

I HAVE already stated that the results of the admirable investigations of Lawes and Gilbert prove that the non-nitro-

* So termed because it is the basis of the common oils ; the fluid portion of fat is composed of oleine.

genous constituents of the carcasses of oxen, sheep, and pigs exceed in weight their nitrogenous elements. This fact is suggestive of many important questions. What relation is there between the composition of an animal and that of its food? Should an animal whose body contains three times as much fat as lean flesh, be supplied with food containing three times as much fat-formers as flesh-formers? To these questions there is some difficulty in replying. There *is* a relationship between the composition of the body of an animal and that of its food; but the relationship varies so greatly that it is impossible to determine with any degree of accuracy the quantity of fat-formers which is required to produce a given weight of fat in animals, taken *in globo*. If, however, we deal with a particular animal placed under certain conditions, it is then possible to ascertain the amount of fat which a given weight of non-plastic food will produce. For the greater part of our knowledge on this point, as on so many others, in the feeding of stock, we are indebted to Lawes and Gilbert. In the case of sheep fed upon fattening food these inquirers found that every 100lbs. of dry* non-nitrogenous substances consumed by them produced, on an average, an increase of 10lbs. in the weight of their fat. In the case of pigs, also, supplied with food, the proportion of non-nitrogenous matters appropriated to the animal's increase was double that so applied in the bodies of the sheep. As the food supplied to these animals contained but a very small proportion of ready-formed fat, it was inferred that four-fifths of the fat of the increase was derived from the sugar, starch, cellulose, and pectine bodies.

These tables exhibit in a condensed form the results of one of the elaborate series of experiments in relation to this point carried out by Lawes and Gilbert:—

---

* The term *dry* is applied to the *solid* constituents of the food. Thus, a pig fed with 100lbs. of potatoes would be said to have been supplied with 25lbs. of dry potatoes, because water forms 75 per cent. of the weight of those tubers.

THE ANIMAL AND ITS FOOD.

ESTIMATED AMOUNT OF CERTAIN CONSTITUENTS STORED UP IN *INCREASE*, FOR 100 PARTS OF EACH CONSUMED IN FOOD BY FATTENING SHEEP.

General Particulars of the Experiments.					Amount of each Class in Increase for 100 of the same consumed in Food.			
Breed.	No. of Animals.	Duration.	Description of Fattening Food.		Mineral matter (ash).*	Nitrogenous compounds (dry).	Non-nitrogenous substance.	Total dry substance.
			Given in limited quantity.	Given ad libitum.				

Class I.

		wks.	dys.						
Cotswolds	46	19	5	Oilcake and clover chaff	Swedish turnips.	3·98	4·43	11·6	9·60
Leicesters	40	20	0			3·15	3·39	12·0	9·48
Cross-bred wethers	40	20	0			3·24	3·60	11·6	9·31
Cross-bred ewes	40	20	0			3·25	3·60	11·8	9·40
Hants Downs	40	26	0			3·40	4·28	10·3	8·49
Sussex Downs	40	26	0			3·30	4·16	10·3	8·44
Means						3·39	3·91	11·3	9·12

Class III.—(Series 1.)

Hants Downs	5	13	6	Oilcake	Swedish turnips.	4·16	4·01	11·1	9·33
	5	13	6	Oats		5·73	7·07	10·0	9·45
	5	13	6	Clover chaff		3·98	7·44	9·0	8·49
Means						4·62	6·17	10·0	9·09

Class IV.—(Series 2.)

Hants Downs	5	19	1	Oilcake	Clover chaff.	1·69	2·20	6·3	5·07
	5	19	1	Linseed		1·81	2·32	6·2	5·19
	5	19	1	Barley		1·75	2·82	5·7	5·00
	5	19	1	Malt		1·46	2·17	5·3	4·61
Means						1·68	2·38	5·9	4·97

* The amounts of "mineral matter" are too high, owing to the adventitious matters (dirt) retained by the wool.

General Particulars of the Experiments.					Amount of each Class in Increase for 100 of the same consumed in Food.			
Breed.	No. of Animals.	Duration.	Description of Fattening* Food.		Mineral matter (ash).*	Nitrogenous compounds (dry).	Non-nitrogenous substance.	Total dry substance.
			Given in limited quantity.	Given ad libitum.				
Class V.—(Series 4.)								
Hants Downs	4	wks. dys. 10 0	Barley ground	Mangolds.	3·80	5·65	9·8	8·91
	5	10 0	Malt, ground, & malt dust.		4·04	6·18	10·4	9·49
	4	10 0	Barley ground and steeped.		3·72	6·35	8·9	8·28
	4	10 0	Malt, ground and steeped, & malt dust.		2·95	4·34	9·3	8·23
	5	10 0	Malt, ground, & malt dust.		3·46	5·46	9·1	8·25
Means					3·59	5·60	9·5	8·63
Means of all					3·27	4·41	9·4	8·06

ESTIMATED AMOUNT OF CERTAIN CONSTITUENTS STORED UP IN *INCREASE*, FOR 100 OF EACH CONSUMED IN FOOD, BY FATTENING PIGS.

General Particulars of the Experiments.					Amount of each Class in Increase for 100 of the same consumed in Food.				
No. of Animals.	Duration.	Description of Fattening Food.			Mineral matter (ash).	Nitrogenous compounds (dry).	Non-nitrogenous substance.	Total dry substance.	Fat.
		Given in limited quantities.		Given ad libitum.					
The Analysed "Fat Pig."†									
1	weeks 10	Mixture of bran 1, bean and lentil-meal 2, and barley-meal 3 parts, ad libitum			2·66	7·76	17·6	14·9	40·5

* The amounts of "mineral matter" are too high, owing to the adventitious matters (dirt) retained by the wool.
† This pig was completely analysed by Lawes and Gilbert.

THE ANIMAL AND ITS FOOD.

				Amount of each Class in Increase for 100 of the same consumed in Food.				
No. of Animals.	Duration.	Description of Fattening Food.		Mineral matter (ash).	Nitrogenous compounds (dry).	Non-nitrogenous substance.	Total dry substance.	Fat.
		Given in limited quantities.	Given ad libitum.					

Series I.

No.	weeks	Limited	Ad libitum	Min	Nitro	Non-nitro	Total	Fat
3	8	None	Bean & lentil-meal	0·68	4·88	25·3	17·5	621
3		Indian-meal		1·86	6·39	23·7	17·9	477
3		Indian-meal and bran		0·33	5·02	21·1	16·1	362
3		None	Indian-meal	2·09	9·28	20·9	18·6	300
3		Bean and lentil-meal		0·99	9·18	20·9	18·4	324
3		Bran		2·35	12·10	20·3	18·7	300
3		Bean, lentil-meal, and bran		2·71	10·03	21·3	18·5	307
3			Bean, lentil-meal, Indian-meal, bran, ad libitum.	0·22	5·65	21·1	16·8	362
		Means		0·74	7·82	21·8	17·8	382

Series II.

No.	weeks	Limited	Ad libitum	Min	Nitro	Non-nitro	Total	Fat
3	8	None	Bean & lentil-meal	3·20	3·12	26·5	18·2	801
3		Barley-meal		0·16	4·65	19·2	14·7	575
3		Bran		0·16	3·99	21·2	15·2	547
3		Barley-meal and bran		0·75	4·57	20·1	15·6	514
3		None	Barley-meal	0·56	10·09	18·5	16·9	574
3		Bean and lentil-meal		0·53	6·57	21·1	17·5	620
3		Bran		0·49	9·79	18·9	16·9	506
3		Bean, lentil-meal, and bran.		4·33	4·49	22·7	18·0	578
6			Mixture of bran 1, barley-meal 2, and bean lentil-meal 3 parts, ad libitum.	0·27	5·65	20·4	16·1	495
6			Mixture of bran 1, bean lentil-meal 2, barley-meal 3 parts, ad libitum.	1·58	8·10	21·1	17·6	515
		Means		0·59	6·10	21·0	16·7	572

Series III.

No.	weeks	Limited	Ad libitum	Min	Nitro	Non-nitro	Total	Fat
4	8	Dried Cod Fish	Bran & Indian-meal (equal parts).	1·06	5·06	24·3	18·1	315
4			Indian-meal	0·26	8·16	25·6	20·9	352
		Means		0·66	6·61	24·9	19·5	333

No. of Animals.	Duration.	Description of Fattening Food.		Mineral matter (ash).	Nitrogenous compounds (dry).	Non-nitrogenous substance.	Total dry substance.	Fat.
		Given in limited quantities.	Given ad libitum.					

General Particulars of the Experiments. — Amount of each Class in Increase for 100 of the same consumed in Food.

Series IV.

	weeks							
3	10	Lentil-meal & bran	Sugar	3·07	9·30	19·4	16·9	
3			Starch	3·18	9·36	19·4	16·9	
3			Sugar & starch	4·06	10·78	17·7	16·1	
3		Lentils, bran, sugar, starch, ad libitum.		4·80	9·96	18·7	16·5	
		Means		3·78	9·85	18·8	16·6	
		Means of all		0·58	7·34	21·2	17·3	472

The larger appropriation of the non-nitrogenous constituents of its food by the pig, as compared with the sheep, must not be attributed solely to its greater tendency to fatten, but partly to the far more digestible nature of the food supplied to it.

SECTION V.

RELATION BETWEEN THE QUANTITY OF FOOD CONSUMED BY AN ANIMAL, AND THE INCREASE IN ITS WEIGHT, OR OF THE AMOUNT OF ITS WORK.

THE manifestations of that wondrous and mysterious principle, *life*, are completely dependent upon the decomposition of organised matter. Not an effort of the mind, not a motion of the body, can be accomplished without involving the destruction of a portion of the tissues. In a general sense we may regard the fat of the animal to be its store of fuel, and its lean flesh to be the source of its motive power. As the evolution of heat within the body is proportionate to the quantity of fat consumed, so also is the amount of force developed in the animal mechanism in a direct ratio to the proportion of flesh decomposed. The quantity of fat burned in the body is estimated by the amount of carbonic acid gas expired from the lungs and perspired through the skin; the proportion of flesh disorganised is ascertained by the quantity of urea eliminated in the liquid egesta. The amount of urea excreted daily by a man is influenced by the activity of his mind, as well as by that of his body. A man engaged in physical labor wears out more of his body than one who does no work; and a man occupied in a pursuit involving intense mental application, consumes a greater proportion of his tissue than the man who works only with his body.* In each of these cases, there is a different amount of tissue disorganised, and consequently a demand for different amounts of food, with which to repair the waste. But all the food consumed by a man is not devoted to the reparation of the tissue worn out in the operations of thinking and working. A human being whose mind is a perfect blank, and who performs no bodily work, excretes a large quantity of urea, the representative of an equivalent amount of worn-out flesh. In fact the greater part of the food consumed by a man serves merely to sustain the functions of the body—the circulation of the blood —the action of the heart—the movements of the muscles concerned in respiration—in a word, the various motions of the body which are independent of the will. According to

* The results of recent and accurately conducted investigations prove that men engaged in occupations requiring the highest exercise of the intellectual faculties, require more nutritious food, and even a greater quantity of nutriment, than the hardest worked laborers, such as paviours, and navvies. I have been assured by an extensive manufacturer, that on promoting his workmen to situations of *greater* responsibility but *less* physically laborious than those previously filled by them, he found that they required more food and that, too, of a better quality. This change in their appetite was not the result of increased wages, which in most cases remained the same—the decrease in the amount of labour exacted being considered in most cases a sufficient equivalent for the increased responsibility thrown upon them.

Professor Haughton, about three-fourths of the food of a working man of 150 lbs. weight, are used in merely keeping him *alive*, the remaining fourth is expended in the production of mechanical force, constituting his daily toil.

In the nutrition of the lower animals, as in that of man, the amount of food made use of by a particular individual depends upon its age, its weight, the amount of work it performs, and probably its temper. As three-fourths of the weight of the food of a laboring man are expended in merely keeping him alive, it is obvious that the withholding of the remaining fourth would render him incapable of working. An amount of food which adequately maintains the vital and mechanical powers of three men, serves merely to keep four alive. It is the same with the horse, the ox, and every other animal useful to man : each makes use of a certain amount of food, *for its own purposes;* all that is consumed beyond that is applied for the benefit of its owner. Let us take the case of two of our most useful quadrupeds—the horse and the ox. The horse is used as an immediate source of motive power. For this purpose food is supplied to it, the greater portion of which is consumed in keeping the animal alive, and the rest for the development of its motive power. Abundance of food is as necessary to the natural mechanism, the horse, as fuel is to the artificial mechanism, the steam-engine. In each case the amount of force developed is, within certain limits, proportionate to the quantity of vegetable or altered vegetable matter consumed. The greater portion of the ox's food is also consumed in keeping its body alive, and the rest, instead of being expended in the development of motive power, accumulates as surplus stores of flesh, which in due time are applied to the purpose of repairing the organisms of men. It is evident then, that the greater sufferer from the deficient supply of food to animals is their owner. That they cannot be *taught* to *fast* is a fact which does not appear very patent to some minds. The man who sought by gradually reducing the daily quantum of his horse's provender to accustom it to work without eating, was justly punished for his ignorant cruelty. The day before the horse's allowance was to be reduced to pure water, and when its owner's hope appeared certain of speedy realisation, the animal died. There are men who act almost as foolishly as the parsimonious horse owner in this fable did; and who are as properly punished as he was. Such men are to be found in the farmers who overstock their sheep pastures, and whose "lean kine" are the *laughing stock* of their more intelligent neighbours.

The weight of a working full-grown horse does not vary from day to day, as the weight of its egesta is equal to that of its food. The desideratum in the case of the working animal is that its food should be as thoroughly decomposed as possible, and the force pent up in it liberated within the animal's body: as an ox, on the contrary, increases in weight from day to day, it is desirable that as little as possible of its food should be disorganised. The wasteful expenditure of the animal's fat may be obviated by shelter, and the application of artificial heat: the retardation of the destruction of its flesh is even more under our control; for, as active muscular exertion involves the decomposition of tissue, we have merely to diminish the activity of the motions which cause this waste. This, in practice, is effected by stall-feeding. Confined within the narrow boundaries of the stall, the muscular action of the animal is reduced to a minimum, or limited to those uncontrollable actions which are conditions in the maintenance of animal life.

The proportion of the food of oxen, sheep, and pigs, which is consumed in maintaining their vital functions, has not been accurately ascertained; probably, as in the case of man, it is strictly proportionate to the animal's weight. We can determine the amount of plastic food consumed by an animal during a given period: we can ascertain the increase (if any) in the weight of its body; and finally, we can weigh and analyse its egesta. With these data it is comparatively easy to ascertain the quantity of food which produced the increase in the animal's weight; but they do not enable us to determine the amount expended in keeping it alive, because the egesta might be largely made up of unappropriated food—organised matter which had done no work in the animal body. When we come to know the precise quantity of nitrogen, in a purely, or nearly pure, mineral form* excreted by an animal, then we shall be in a position to estimate the proportion of its food expended in sustaining the essential vital processes which continuously go on in its body. But although we are in ignorance as to the precise quantity of flesh-formers expended in keeping the animal alive, we know pretty accurately the amount which is consumed in producing a given weight of its flesh, or rather in causing a certain increase in its weight. This knowledge is the result of numerous investigations, of which by far the most valuable are those of Lawes and Gilbert. These experimenters found that fattening pigs stored up about 7½ per cent. of the plastic materials of their food, whilst sheep accumulated somewhat less than 5 per cent. That is, 92½ out of every 100 lbs. weight of the nitrogenous food of the pig, and 95 out of every 100 lbs. of that of the sheep, are eliminated in the excretions of those animals.

It appears from the results of Lawes and Gilbert's experiments, that pigs store up in their *increase* about 20 per cent., sheep 12 per cent., and oxen 8 per cent. of their (dry) food. The relative increase of the fatty, nitrogenous, and mineral constituents whilst fattening, are shown in this table.

Cases.	Estimated per cent. in Increase whilst Fattening.			
	Mineral matter (ash).	Nitrogenous matter (dry).	Fat (dry).	Total dry substance.
Average of 98 oxen	1·47	7·69	66·2	75·4
Average of 348 sheep	1·80	7·13	70·4	79·53
Average of 80 pigs	0·44	6·44	71·5	78·40

* As ammonia, urea, uric acid, or hippuric acid; all of which are nearly or perfectly mineralised subtances.

The quantity of food consumed daily by an animal is, as might be expected, proportionate to the weight of its body. The pig consumes, for every 100 lbs. of its weight, from 26 to 30 lbs. of food, the sheep 15 lbs., and the ox 12 to 13 lbs. These figures and the statements which I have made relative to the proportions of fat and plastic elements in the animals' bodies, apply to them in their fattening state, and when the food is of a highly nutritious character. The calf and the young pig will make use—to cause their increase—of a larger portion of nitrogenous matters. The sheep, however, being early brought to maturity, will, even when very young, store up the plastic and non-plastic constituents of its food, in nearly the same relative proportions that I have mentioned.

As it is the food taken into the body that produces heat and motion, it might at first sight appear an easy matter to determine the amount of heat or of motion which a given weight of a particular kind of food is capable of producing within the animal mechanism. But this performance is not so easy a task as it appears to be. In the first place, all of the food may not be perfectly oxidised, though thoroughly disorganised within the body; secondly, as animals rarely subsist on one kind of food, it is difficult, when they are supplied with mixed aliments, to determine which of them is the most perfectly decomposed. But though the difficulties which I have mentioned, and many others, render the task of determining the nutritive values of food substances difficult, the problem is by no means insoluble, and, in fact, is in a fair way of being solved. Professor Frankland, in a paper published in the number of the *Philosophical Magazine* for September, 1866, determines the relative alimental value of foods by ascertaining the quantity of heat evolved by each when burned in oxygen gas. From the results of these researches he has constructed a table, showing the amount of food necessary to keep a man alive for twenty-four hours. The following figures, which I select from this table, are of interest to the stock-feeder:—

Kinds of Food.	Weight necessary to sustain a man's life for twenty-four hours. Ounces.
Potatoes	13·4
Apples	20·7
Oatmeal	3·4
Flour	3·5
Pea Meal	3·5
Bread	6·4
Milk	21·2
Carrots	25·6
Cabbage	31·8
Butter	1·8
Lump Sugar	3·9

These figures show the relative calefacient, or heat-producing powers of the different foods named *outside* the body; but there is some doubt as to their having the same relative values when burned *within* the body. The woody fibre of the carrots and cabbages is very combustible in the coal furnace, but it is very doubtful if more than 20 or 30 per cent. of this substance is ever burned in the *animal furnace*. However, such inquiries as those carried out by Frankland possess great value; and tables constructed upon their results cannot fail to be useful in the drawing up of dietary scales, whether for man or for the inferior animals.

I may here remark, that in my opinion the nutritive value of food admits of being very accurately determined by the adoption of the following method :—

1. The animal experimented upon to be supplied daily with a weighed quantity of food, the composition and calefacient value of which had been accurately determined. 2. The gases, vapors, and liquid and solid egesta thrown off from its body to be collected, analysed, and the calefacient* value of the combustible portion of them to be

---

* The excrements of animals are capable of evolving, by combustion, enormous amounts of heat.

determined. 3. The increase (if any) of the weight of the animal to be ascertained. 4. The difference between the amount of heat evolvable by the foods before being consumed, and that actually obtained by the combustion of the egesta into which they were ultimately converted, would be the amount actually set free and rendered available within the body. The calculations would be somewhat affected by an increase in the weight of the animal's body; but it would not be difficult to keep the weight stationary, or nearly so, and there are other ways of getting over such a difficulty. An experiment such as this would be a costly one, and could not be properly conducted unless by the aid of an apparatus similar to that employed by Pettenkofer in his experiments on respiration. This apparatus, which was made at the expense of the King of Bavaria, cost nearly £600.

Value of Manure.—It is a complication in the question of the economic feeding of the farm animals that the value of their manure must be taken into account. Of the three classes of food constituents, two—the mineral and nitrogenous—are recoverable in the animal's body and manure; the non-nitrogenous is partly recoverable in the fat. I shall take the case of a sheep, which will consume weekly per 100 lbs. of its weight, 12 lbs. of fat-formers, and 3 lbs. of flesh-formers. Twelve per cent. of the fat-formers will be retained in the *increase*, but the rest will be expended in keeping the animal warm, and the products of its combustion—carbonic acid and water—will be useless to the farmer. It is, therefore, desirable to diminish as much as possible the combustion of fatty matter in the animal's body; and this is effected, as I have already explained, by keeping it in a warm place. Of the flesh-forming substance only five per cent. is retained in the increase, the rest is partly consumed in carrying on the movements of the animal—partly expelled from its body unaltered, or but slightly altered, in composition. The solid excrement of the animal contains all the undigested food; but of this only the mineral and nitrogenous constituents are valuable as manure. The nitrogen of the plastic materials which are expended in maintaining the functions of the body is eliminated from the lungs, through the skin, and by the kidneys—perhaps also, but certainly only to a small extent, by the rectum.

The food consumed by an animal is disposed of in the following way :—A portion passes unchanged, or but slightly altered, through the body; another part is assimilated and subsequently disorganised and ejected; the rest is converted into the carcass of the animal at the time of its death. The undigested food and aliment which had undergone conversion into flesh and other tissues, and subsequent disorganisation, constitute the excrements, or manure, of the animal. The richer in nitrogen and phosphoric acid the food is, the more valuable will be the manure; so that the money value of a feeding stuff is not determinable merely by the amount of flesh which it makes, but also, and to a great extent, by the value of the manure into which it is ultimately converted.

Corn and oil-cakes are powerful fertilisers of the soil; but the three principles which constitute their manurial value——namely, nitrogen (ammonia), phosphoric acid, and potash—are purchasable at far lower prices in guano and other manures. Nevertheless, many farmers believe that the most economical way to produce good manure is to feed their stock with concentrated aliment, in order to greatly increase the value of their excreta. They consider that a pound's worth of oil-cake, or of corn, will produce at least a pound's worth of meat, and that the manure will be had for nothing, or, rather, will be the profit of the business. The richer food is in nitrogen and phosphoric acid, the more valuable will be the manure it yields. It follows, therefore, that if two kinds of feeding stuff produce equal amounts of meat, that the preference should be given to that which contains the more nitrogen and phosphoric acid. Mr. Lawes, who has thrown light upon this point, as well as upon so many others, has made careful estimates of the value of the manure produced from different foods. They are given in the following table:—

TABLE

Showing the estimated value of the manure obtained on the consumption of one ton of different articles of food; each supposed to be of good quality of its kind.

Description of Food.	Estimated Money Value of the Manure from One Ton of each Food.
1. Decorticated cotton-seed cake	£6 10 0
2. Rape-cake	4 18 0
3. Linseed-cake	4 12 0
4. Malt-dust	4 5 0
5. Lentils	3 17 0
6. Linseed	3 13 0
7. Tares	3 13 6
8. Beans	3 13 6
9. Peas	3 2 6
10. Locust beans	1 2 6 (?)
11. Oats	1 14 6
12. Wheat	1 13 0
13. Indian corn	1 11 6
14. Malt	1 11 6
15. Barley	1 9 6
16. Clover-hay	2 5 0
17. Meadow-hay	1 10 0
18. Oat-straw	0 13 6
19. Wheat-straw	0 12 6
20. Barley-straw	0 10 6
21. Potatoes	0 7 0
22. Mangolds	0 5 0
23. Swedish turnips	0 4 3
24. Common turnips	0 4 0
25. Carrots	0 4 0

All the saline matter contained in the food is either converted into flesh, or is recoverable in the form of manure, but a portion of its nitrogen appears to be lost by respiration and perspiration. Reiset states that 100 parts of the nitrogen of food given to sheep upon which he experimented, were disposed of as follows:—

Recovered in the excreta	58·3
Recovered in the meat, tallow, and skin	13·7
Lost in respiration	28·0
	100·00

Haughton's experiments, performed upon men, gave results which proved that no portion of the nitrogen of their food was lost by perspiration or by respiration. Barral, on the contrary, asserts that nitrogen is given off from the bodies of both man and the inferior animals. Boussingault states that horses, sheep, and pigs exhale nitrogen. A cow, giving milk, on which he had experimented, lost 15 per cent. of the nitrogen of its food by perspiration. The amount of nitrogen which Reiset states that sheep exhale is exceedingly great, and it is difficult to reconcile his results with those obtained by Voit, Bischoff, Regnault, Pettenkofer, and Haughton. Of course, men and sheep are widely different animals; but still it is unlikely that all the nitrogen of the food of man should be recoverable in his egesta, whilst nearly a third of the nitrogen of the food of the sheep should be dissipated as gas. I think further experiments are necessary before this point can be regarded as settled; and it is probable that it will yet be found that all, or nearly all, of the nitrogen of the food of animals is recoverable in their egesta.

Regarding, then, an animal as a mechanism by which meat is to be "manufactured," five economic points in relation to it demand the feeder's attention: these are—the first cost of the mechanism, the expense of maintaining the mechanism in working order, the price of the raw materials intended for conversion into meat, the value of the meat, and the value of the manure. In proportion to the attention given to these points, will be the feeder's profits; but they are, to some extent, affected by the climatic, geographic, and other conditions under which the farm is placed.

PART II.

ON THE BREEDING AND BREEDS OF STOCK.

SECTION I.

THE BREEDING OF STOCK.

Cross Breeding.—For many years past feeders have zealously occupied themselves in the improvement of their stock, and the result of their labors is observable in the marked superiority of the breeds of the present day over their ancestors in the last century. The improvement of animals designed as food for man is effected by keeping them on a liberal dietary, by selecting only the best individuals for sires and dams, and by combining the excellencies of two or more varieties of a species in one breed. A species consists of a number of animals which exhibit so many points of resemblance, that they are regarded by the great majority of naturalists to be the descendants of a single pair. If we except the believers in the hypotheses relative to the origin of existing varieties of animals and plants, propounded by Lamarck, Darwin, and other naturalists of the "advanced school," there is a general belief in the immutability of species. The individuals of an existing species, say dogs, can never acquire the peculiar features of another species; nor can their descendants, if we except hybrids, ever become animals in which the characteristics of the dog tribe are irrecognisable. By various influences, such as, for example, differences in food and climate, and domestication, a species may be split into *varieties*, or *breeds*, all of which, however, retain the more important characteristics of the primordial type. There appears to be no limit to the varieties of dogs, yet one can perceive by a glance that there is no specific difference between the huge Mont St. Bernard dog and the diminutive poodle, or between the sparse greyhound and the burly mastiff. All the varieties of our domestic fowl have been traced to a common origin—the wild Indian fowl (*Gallus bankiva*). Even Darwin admits that all the existing kinds of horses are, in all probability, the descendants of an original stock; and it is generally agreed that the scores of varieties of pigeons own a common ancestor in the rock pigeon (*Columba livia*).

As certain individuals are grouped by naturalists into species, so particular species, which in habits and general appearance resemble each other, are arranged under the head of genus. The horse, the ass, and the zebra are formed on nearly the same anatomical plan; they are therefore classed together, and designated the genus *Equus*, a term derived from the Latin word *equus*, a horse—that animal being regarded as the type, or perfect member of the group. Thus the horse, in the nomenclature of the naturalist, is termed *Equus caballus;* the ass, *Equus asinus;* and the zebra, *Equus zebra*. By a further extension of this principle of classification, very closely allied genera are united under the term of *family*.

The different varieties of the same species breed, as might be anticipated, freely together; but it frequently happens that two individuals of different species pair, and produce an animal which inherits some of the properties of each of its progenitors. These half-breeds are termed *hybrids*, or *mules*, and we have familiar examples of them in the common mule and the jennet. As a general rule, animals exhibit a disinclination to breed with other than members of their own species; and although the interference of man may overcome this natural repugnance, he can only effect the fruitful congress of individuals belonging to closely allied species, being members of the same genus. Hybrids in the genus *Equus* are very common. A cross has been produced between the he-goat and the ewe; the camel and the dromedary have bred together; and Buffon succeeded in producing a hybrid in which three animals were represented—namely, the bison, the zebu, and the ox. On the other hand, attempts to effect a cross between animals belonging to different families have generally failed; nor is it at all probable that a cross will ever be produced between the pig and the sheep, between the horse and the cow, or, most unlikely of all, between the dog and the cat.

It is the general belief that hybrids are sterile, or, at least, that they are incapable of propagation *inter se*. This may be true with respect to the hybrids of species not very closely allied; but that there are exceptions to the rule is quite clear from Roux's experiments with hares and rabbits. This gentleman, who is, or was, the president of a French agricultural society, but who makes no profession of scientific knowledge, has succeeded, after several failures, in producing a fruitful cross between the rabbit and the hare. This hybrid has received the name of leporide (from the Latin *leporinus*, pertaining to a hare), and it is different from former crosses, in being five parts hare, and three parts rabbit. M. Roux has bred this hybrid during the last eighteen years, and has not observed the slightest appearance of decay of race manifest itself up to the present, so that, for all practical purposes, the leporide may be regarded as an addition to the distinct species of animals. The leporide fattens rapidly, and with but little expenditure of food. Sold at the age of four months, it realises, in France, a price four times greater than that commanded by a rabbit of the same age; and at a year old it weighs on an average ten pounds, and sometimes as much as sixteen pounds. It breeds at four months, continues thirty days in gestation, and yearly produces five or six litters of from five to eight young. To produce this hybrid is by no means difficult. A leveret, just old enough to dispense with the maternal nutriment, should be placed with a few doe rabbits of his own age, apart from other animals. He will soon become familiar with the does, and when they attain the age of puberty, all the rabbits save one or two should be removed. Speedily those left with the hare will become with young, upon which they should be removed, and replaced by others. After this the hare should be kept in a hutch by himself, and a doe left with him at night only. As the hare is naturally a very shy animal, it will only breed when perfect quietness prevails. The half-bred produced in the first instance should now be put to the hare, and a cross, three parts hare, and one part rabbit, obtained. The permanent breed should then be obtained by crossing the quadroon doe leporide, if I may use the term, with the half-bred buck.

I have directed attention to the production of the leporide because I believe that the problems in relation to it, which have been solved by M. Roux, have an important bearing upon the breeding of animals of greater importance than hares and rabbits. Here we find a race of animals produced by the fusion of two species, which naturally exist in a state of mutual enmity, and which differ in many important respects. The hare and the rabbit are respectively of but little value as food, at least they are of no importance to the feeder; yet a cross between them turns out to be an excellent meat-producing animal, which may be reared with considerable profit to the feeder. It is thus clearly shown that two kinds of animals, neither of which is of great utility, may give rise to an excellent cross, if their blood, so to speak, be blended in proper proportions. A half-bred animal may be less valuable than its parents, but a quadroon may greatly excel its progenitors. The goat and sheep are so closely related that they are classed by naturalists under one head—*Capridæ*. Some kinds of sheep have hair like goats, and certain varieties of goats have fleeces that closely resemble those on the sheep. There are sheep with horns, and goats without those striking appendages. The Cape of Good Hope goat might easily be mistaken for a sheep. It would seem, judging by the results of Roux's experiments, that there is no great difficulty in the way of obtaining a cross between the sheep and the goat. I do not mean an ordinary half-breed, but a prolific hybrid similar to the leporide. Of course, it is impossible, *a priori*, to say whether or not such a hybrid race, supposing it produceable, would be valuable; but as goats can find a subsistence on mountains where sheep would starve, it is possible that an animal, essentially a sheep, but with a streak of goat blood in it, could be profitably kept on very poor uplands. Whether a race of what we might term *caprides* be formed or not we have derived most suggestive information from M. Roux's experiments, which I hope may be turned to account in what is by far the most important field of enquiry, the judicious crossing of varieties of the same species.

It is a *quæstio vexata* whether or not the parents generally exercise different influences upon the shape and size of their offspring. Mr. Spooner supports the supposition—a very popular one—that the sire gives shape to the external organs, whilst the dam affects the internal organisation. I have considerable doubt as to the probability of this theory. The children who spring from the union of a white man with a negress possess physical and intellectual qualities which are nearly if not quite the *mean* of their parents; but the offspring of parents, both of the same race—be it Caucasian, Mongolian, or Indian— frequently conform, intellectually and corporeally, to either of their progenitors. Thus, of the children of a tall, thin, dark man, and a short, fat, fair woman, some will be like their father, and the others will resemble their mother, or, perhaps, all may "take after" either parent. Sometimes a child appears to be in every respect unlike its parents, and occasionally the likeness of an ancestor appears in a descendant, in whom no resemblance to his immediate progenitors can be detected. It is highly probable that both parents exercise, under most circumstances, a joint influence upon the qualities of their offspring, but that one of them may produce so much greater an effect that the influence of the other is not recognisable, except perhaps to a very close observer. But I doubt very much that any particular organ of the offspring is, as a rule, more liable to the influence of the sire than of the dam, or *vice versâ ;* and the breeder who believes that the sire alone is concerned in moulding the external form of the offspring, and who consequently pays no attention to this point in the dam, will often find himself out in his reckonings. In order to be certain of a satisfactory result, the dam should in every respect be equal to the sire. In practice, however, this is not always the case, for as sires are so few as compared with the number of dams, the greatest efforts have been directed towards the improvement of the former.

There is, or ought to be, a familiar maxim with breeders, that "like begets like, or the likeness of an ancestor." This is a "wise saw," of which there are many "modern instances :" the excellencies or defects of sire or dam are certain to be transmitted through several generations, though they may not appear in all. As a general rule, good animals will produce a good, and defective animals a defective, offspring, but it sometimes happens that a bull or cow, of the best blood, is decidedly inferior, whilst really good animals are occasionally the produce of parents of "low degree." If the defects or excellencies of animals were ineradicable there would be no need for the science of breeding; but by the continual selection of only the most superior animals for breeding purposes the defects of a species gradually disappear, and the good qualities are alone transmitted. As, however, animals that are used as food for man are to some extent in an abnormal condition, the points which may be excellencies in that state, would not have been such in the original condition of the animal. We find, therefore, that the improved breeds of oxen and sheep exhibit some tendency to revert to their original condition, and it is only by close attention to the diet, breeding, and general management of these animals that this tendency can be successfully resisted. Sometimes, however, an animal of even the best breed will "return to nature," or will acquire some undesirable quality; such an animal should be rejected for breeding purposes, for its defects would in all probability be transmitted to its descendants, near or remote. A case, which admirably illustrates this point, is recorded in the *Philosophical Transactions* for 1813, and it is sufficiently interesting to be mentioned here :—

Seth Wright, who possessed a small farm on the Charles River, about sixteen miles from Boston, had a small flock, consisting of fifteen ewes and one ram. One of these ewes, in 1791, produced a singular-shaped male lamb. Wright was advised to kill his former ram and keep this new one in place of it; the consequence was, the formation of a new breed of sheep, which gradually spread over a considerable part of New England, but the introduction of the Merino has nearly destroyed them again. This new variety was called the Otter, or "Ankon" breed. They are remarkable for the shortness of their legs, and the crookedness of their forelegs, like an elbow. They are much more feeble and much smaller than the common sheep, and less able to break over low fences; and this was the reason of their being continued and propagated.

Here we have an instance of an animal propagating a defect through a great number of descendants, though it had not acquired it from its own ancestors. It is, however, probable that occasionally a male descendant of this short-legged ram possessed considerably longer organs of locomotion than the founder of his breed; and, consequently, if selected for breeding purposes might become the founder of a long-legged variety, in which, however, a couple of pairs of short-legs would occasionally present themselves. I have a notion that the higher animals are in the scale of being, the greater is their tendency to transmit their acquired good or bad habits to their posterity. Dogs are, perhaps, the most intelligent of the inferior animals, and it is well known that they transmit to their offspring their acquired as well as their natural habits. I doubt very much that those most stupid of creatures, guinea-pigs, possess this property in any sensible degree; or, indeed, that like the canine tribe, they can be readily made to acquire artificial peculiarities: but there once flourished a "learned pig," and it would be worth inquiring whether or not its descendants, like the descendants of the trained setter, and pointer, were at all benefited by the education of their ancestor. I shall conclude this part of my subject in the words of Professor Tanner: "In all cases where the breed has been carefully preserved pure, great benefit will result from doing so. The character of a breed becomes more and more concentrated and confirmed in a pedigree animal, and this character is rendered more fully hereditary in proportion to the number of generations through which it has been transmitted. By the aid of pedigree, purity of blood may be insured, and a systematic plan adopted by which we can perpetuate distinct families, and thereby obtain a change of blood without its being a cross. It is evident that any one adopting a systematic arrangement will be able to do this more effectually than another without this aid. This is the more important when the number of families is small, as is the case with Devons and Herefords, especially the former. The individual animals from which the Devons are descended are very limited in number, and in a few hands; but, with some honourable exceptions, little attention is given to this point. The importance is rendered evident by the decreasing size of the breed, the number of barren heifers, and the increased delicacy of constitution shown in the stock of many breeders of that district who are not particular in this respect. The contrast between such herds, and those in which more care and judgment are exercised, renders the advantages of attention to pedigree very evident; for here the strength of constitution is retained, together with many of the advantages of this valuable breed."

SECTION II.

THE BREEDS OF STOCK.

THE nature of the animal determines, as I have already stated, the proportion of its food carried off in its increase; but this point is also greatly influenced by its *variety*, or *breed.* Certain breeds which have for a long period been kept on bulky food, and obliged to roam in quest of it, appear to have acquired a normal tendency to *leanness.* No doubt, if they were supplied with highly nutritious food for many successive generations, these breeds might eventually exhibit as great a tendency to fatten as they now do to remain in a lean condition. As it is, the horned cattle of Kerry, Wales, and some other regions, rarely become fat, no matter how abundantly they may be supplied with fattening food. On the other hand, the Herefords, but more especially the Shorthorns, exhibit a natural disposition to obesity, and such animals alone should be stall-fed. It is noteworthy that animals which are naturally disposed to yield abundance of milk are often the best adapted for fattening; but it would appear that the continuous use of highly fattening food, and the observance of the various other conditions in the *forcing* system, diminish the activity of the lacteal secretion, and increase the tendency to fatness in the races of the bovine tribe. The Shorthorns were at one time famous for their milking capabilities, but latterly their galactophoric reputation has greatly declined. Still I am disposed to believe, that if some of those animals were placed under conditions favorable to the improvement of dairy stock, herds of Shorthorn milch cows could be obtained which would vie in their own line with the famous fat-disposed oxen of the same breed.

In sheep the tendency to early maturity and to fatten is greatly influenced by the breed. The Leicester, even when kept on inferior pasture, fattens so rapidly that in eighteen months it is fit for the butcher; whilst the Merino, though supplied with excellent herbage, must be preserved for nearly four years before it is ready for the shambles. The crossing of good herds has resulted in the development of numerous varieties, all remarkable for their aptitude to fatten and to arrive early at maturity. The Leicester—itself supposed to be a cross—has greatly improved the Lincoln, and the Hampshire and Southdown have produced an excellent cross. Of course, each breed and cross has its admirers; indeed, the differences of opinion which prevail in relation to the relative merits of the Lincoln and the Leicester—the Southdown and the Shropshiredown—the Dorset and the Somerset—occasionally culminate into newspaper controversies of an exceedingly ascerb character. There is no doubt but that particular breeds of sheep thrive in localities and under conditions which are inimical to other varieties; but still it is equally evident that, *cæteris paribus*, one kind of sheep will store up in its increase a larger proportion of its food than another kind, and will arrive earlier at maturity. It is the knowledge of this fact which has led to the great estimation in which are held some half-dozen out of the numerous breeds and cross-breeds of that animal. In 1861 an interesting experiment was made by the Parlington Farmers' Club with the object of testing the relative merits of several varieties of sheep. The results are shown in the tables:—

TABLE I.

Description of Class of Sheep.	Live Weight of Six Wethers when Shorn, 26th February, 1862.		Weight of Mutton when Slaughtered.		Weight of Tallow.	Weight of Wool.	Weight of Pelts.	Weights gained during the time of Feeding from the 11th November, 1861, to 14th February, 1862.					
								In Live Weight.		In Mutton.		In Wool.	
	st.	lb.	st.	lb.	lb.	lb.	lb.	st.	lb.	st.	lb.	lb.	oz.
Cross from the Teeswater	85	3	53	1	106	43	85	13	7	8	6	14	5
North Sheep	83	12	53	12	96	43½	83	12	11	8	3	14	8
Lincolns	92	1	59	12	105	66	103	16	1	10	7	22	0
South Downs	71	0	47	7	97¼	28	65¾	11	13	8	0	9	5
Shropshire Downs	85	6	53	1	103	42½	91	15	11	9	12	14	3
Leicesters	80	9	53	4	90¼	44	78½	14	10	9	10	14	11
Cotswolds	76	5	47	6	79	54	90	12	6	7	11	18	0

TABLE II.

Description of Sheep.	Value of the preceding Mutton and Wool so gained.								Food consumed during time of Feeding.		Value of the Food, calculating Turnips at 6s. 8d., and Cake at £10 10s. per ton.*			Value of the Mutton and Wool.			Value of Food deducted from Value of Mutton and Wool, showing real value of the different Sheep.		
	Price of the Mutton.				Price of the Wool.														
	p.lb				p.lb.				Swd. Tnp.	Lnd. Cke.									
	d.	£	s.	d.	d.	£	s.	d.	st.	lb.	£	s.	d.	£	s.	d.	£	s.	d.
Teeswater, Cross ...	6	2	19	0	18	1	1	6	978	300	3	8	10½	4	0	6	0	11	7
North Shropshire	6	2	17	6	17½	1	1	1¾	914	300	3	6	2½	3	18	7¾	0	12	5
*Lincolnshire	5¾	3	10	5¼	18	1	13	0	936	363	3	13	0¼	5	3	5¼	1	10	5
Southdowns ..	6½	3	0	8	17	0	13	2½	684	300	2	16	7½	3	13	10½	0	17	3
Shropshire ...	6¼	3	11	10½	17½	1	0	7¾	924	300	3	6	7¾	4	12	6¼	1	5	10
Leicester ...	5¾	3	5	2	18	1	2	0	877	300	3	4	8	4	7	2	1	2	6
Cotswolds ..	6	2	14	6	18	1	7	0	926	300	3	6	8½	4	1	6	0	14	9½

These results, taken with the customary *grain of salt*, tell well for the improved Lincoln; they also clearly show the aptitude to fatten, without much loss in offal, of the Leicester;† and they commend to the lover of good mutton the Shropshire and South-Downs.

In the sixteenth volume of the Journal of the Royal Agricultural Society of England, Mr. Lawes gives some valuable information relative to the comparative fattening qualities of different breeds of sheep. The following table, on this author's authority, shows the average food consumed in producing 100 lbs. increase in live weight:—

* Improved by Leicester blood.

† The object of the first breeders of the Leicester was to produce a sheep which would yield a great carcass, and small offal weight. So far as the results of these experiments go, I think the idea of the founder of this breed has been realised.

Breed.	Oil Cake.	Clover.	Swedes.
Sussex	297¼	285½	3·835¾
Hampshire	291½	261¼	3·966¾
Cross-bred Wethers ...	264½	251¾	3·725¼
Do. Ewes	263½	250¼	3·671
Leicesters	263¾	251¼	3·761
Cotswolds	253½	216¾	3·557½

Some breeds are profitably kept in certain localities, where other kinds would not pay so well: for example, the Devons, according to Mr. Smith, are better adapted than larger breeds for "converting the produce of cold and hilly pastures into meat." It is remarkable that nearly all the best existing breeds of oxen and sheep are crosses. Major Rudd states that the dam of Hubback, the famous founder of pure improved Shorthorns, owed her propensity to fatten to an admixture of Kyloe blood, and also that the sire of Hubback had a stain of Alderney, or Normandy blood. Although the Rudd account of the ancestry of Hubback is not accepted by all the historians of this splendid breed of cattle, there is no doubt but that the breed owes its origin as much to judicious crossing as to careful selection of sires and dams. It must not, however, be imagined that there are no good pure races of stock. There is a perfectly pure, but now scarce, tribe of Kerry oxen, admirably adapted to poor uplands. The excellent Southdown sheep, though in every respect immensely superior to their ancestors in the last century, have not attained to their present superior state by crossing. The high value placed by breeders upon good sires and dams in the approved breeds of stock is shown by the large sums which they frequently realise at sales, or when the former are let out for service. Bakewell received in one season for the use of a ram 400 guineas each from two breeders, and they did not retain the animal during the whole season. Several hundred guineas have lately been more than once paid for a celebrated tup. Colonel Towneley's Shorthorn bull, Master Butterfly, was, not long since, disposed of to an

Australian buyer for £1,260. At the sale of Mr. Bates's stock in 1850, a stock of Shorthorns, including calves, brought on the average £116 5s. per head. At the Earl Ducie's sale in 1852, a three year old cow — Duchess — realised 700 guineas.

The color of an animal is, to some extent, a criterion of the purity of its breed. Roan is a favourite hue with the breeders of Shorthorns. There have been celebrated sires and dams of that breed perfectly white; but that color, or rather absence of color, is now somewhat unpopular, partly from the idea that it is a sign of weakness of constitution—a notion for which there appears to me to be no foundation in fact. The slightest spot of black, or even a very dark shade, is regarded to be a blemish of the most serious kind when observed on the pelt of a Shorthorn. The Herefords are partly white, partly red; the Devon possesses in general a deep red hue; the Suffolks are usually of a dun or faint reddish tint; the Ayrshires are commonly spotted white and red; and the Kerrys are seen in every shade between a jet black and a deep red. Uniformity in color would be most desirable in the case of each variety, and this object could easily be attained if breeders devoted some attention to it.

The Form of Animals.—The functions of an animal are arranged by Bichat, an eminent physiologist, into two classes —those relating to its nutrition, and those exhibited by its muscular and mental systems. The first class of functions comprise the *vegetative*, or organic life of the animal, and the second class constitute its *relative* life. Adopting this arrangement, we may say, then, that those animals in which the vegetative life is far more energetic than the relative life are best suited for the purposes of the feeder. In tigers, wolves, and dogs the relative life predominates over the vegetative; the muscles are almost constantly in a high degree of tension, and the processes of nutrition are in constant requisition to supply the waste of muscle. On the other hand, in oxen, sheep, and pigs, at least when in a state of domesticity, the muscles are not highly developed; they do not largely tax the vegetative processes, and, consequently, the substances elaborated under the influence of the vegetative life rapidly increase. The form of an animal is therefore mainly determined by the activity of its relative life. In a greyhound, the nervous power of which is highly developed, the muscles are large and well-knit, the stomach, intended for the reception of concentrated nutriment only, is small, and the lungs are exceedingly capacious. In such an animal the arrangements for the rapid expenditure of nervous power must be perfect. It is not merely necessary that its muscles should be large and powerful, its lungs must also admit of deep inspirations of oxygen, whereby the motive power wielded by these muscles may be rapidly generated. Now, an animal exactly opposite in organisation to the greyhound would, according to theory, be just the kind to select for the production of meat. The greyhound and the horse expend all their food in the production of motive power; the ox and the sheep, being endowed with but a feeble muscular organisation, use a smaller proportion of their food for carrying on the functions of their relative life, consequently, the weight of their bodies is augmented by the surplus nutriment. It is clear, then, that an animal of a lymphatic temperament, an indolent disposition, a low degree of nervous power, and a tendency to rapid growth, is the *beau ideal* of a "meat-manufacturing machine." Now, as the larger the lungs of an animal are, the greater is its capacity for "burning," or consuming its tissues, one might suppose that small lungs would be a *desideratum* in an ox, or other animal destined for the shambles. This appears to be Liebig's opinion, for in one of his books he states that "a narrow chest (small lungs) is considered by experienced agriculturists a sure sign, in pigs, for example, of easy fattening; and the same remark applies to cows, in reference to the produce of milk—that is, of butter." On this subject Professor Tanner makes the following remarks,

in his excellent Essay on Breeding and Rearing Cattle:*—
"In our high-bred animals we find a small liver and a small
lung, accompanied with a gentle and peaceful disposition.
Now, these conditions, which are so desirable for producing
fat, are equally favorable for yielding butter. The diminished
organs economise the consumption of the carbonaceous matters
in the blood, hence, more remains for conversion into fat, but
equally prepared for yielding cream, if the tendency of the
animal is equally favorable to the same." One would imagine,
from the foregoing passage, that Mr. Tanner and Baron Liebig
coincided in believing small lungs necessary to rapid fattening;
but in another part of his essay, Tanner thus describes one of
the points indicative of a tendency to fatten early:—"The
chest should he bold and prominent, wide and deep, furnished
with a deep but not coarse dewlap." On comparing the two
passages which I have quoted from Tanner's essay, a con-
tradiction is apparent. Mr. Bowly, Major Rudd, and other
eminent breeders and feeders, appear to regard a capacious
chest as the best sign of a fattening property which an animal
could show. Lawes and Gilbert have recorded the weights
of the viscera of a number of animals which, though supplied
with equal quantities of the same kind of food, attained to
different degrees of fatness. On carefully scrutinising these
records, I failed to perceive any constant relation between
the weight of their lungs and their tendency to fatten rapidly.
Some animals with large lungs converted a larger proportion
of their food into meat than others with smaller respiratory
organs, and *vice versâ*. In a state of nature, there is no doubt
but that the lungs of the ox and of the sheep are moderately
large; and it is evident that in their case, as well as in that
of man, over-feeding and confinement tend to diminish their
muscular energy, and, of course, to decrease the capacity of
the lungs. That such a practice does not tend to the improve-

* "Transactions of the Highland and Agricultural Society of Scotland," for July, 1860.

ment of the health of an animal is perfectly evident, but then the perfect ox of nature is very different from the perfect ox of man. The latter is a wide departure from the original type of its species: any marked development of its nervous system is undesirable; and it is valuable in proportion as its purely vegetative functions are most strongly manifested. A young bullock, therefore, of this kind would, no doubt, be the most economical kind to rear, provided that it was perfectly healthy, and capable of assimilating the liberal amount of food supplied to it. But it rarely happens that a young animal with a weakly chest turns out other than a scrofulous or otherwise diseased adult. On the whole, then, I am disposed to believe that whilst naturally small-lunged species may be more prone to fatten than large-chested ones, it is not the case that small-chested individuals fatten more rapidly than larger lunged individuals of the same kind.

The conditions under which oxen, sheep, and pigs have been so long maintained in civilised countries, must have diminished the capacity of their chests in relation to other parts of their bodies; and it may be fairly doubted if any good could result by reducing to still smaller dimensions those most important organs. Probably the lungs and hearts of the improved breeds of stock are already too small, and that it is only the individuals which are least affected in this respect that answer to Mr. Bowly's description of a fat-disposed beast. Whether or not small lungs are desirable in a bullock or milch cow, it is certain that a ram or a bull should be possessed of a capacious chest, for otherwise he will have but little vigour, and will be likely to produce a weakly offspring. A sire should be a perfectly developed animal in every respect—sound lungs and heart, and not over fat. It is sufficient that it belongs to a good fattening breed; but to produce offspring with a tendency to fatness and early maturity, it is not necessary that the sire should himself be obese. It is to be regretted that so many sires of the Shorthorns and other improved varieties should be used for breeding purposes, when their hearts and lungs have become, by over-feeding the animals, unfitted for the proper discharge of their function. The progeny of such sires must *naturally* inherit the *acquired taint* of their diseased progenitors, and prove weakly and unhealthy animals.

With respect to the general outline structure of a bull, he should have a small, well-set head, rounded ribs, straight legs, small bones, and sound internal organs. The following are considered to be the best points in a Shorthorn bull :—A short and moderately small head, with tapering muzzle and broad forehead, furnished with short, white, curved, graceful looking horns; bright, yet mild, large eyes, placed in prominent orbits; dilated nostrils, and flesh-colored nose, and long, thin ears. The neck should be broad, deep, and muscular, sloping in a graceful line from the shoulder to the head. The chest should be wide, deep, projecting, but level in front. The shoulders should be oblique, the blades well set in towards the ribs. The forelegs should be stout, muscular above the knee, and slender below it; the hind legs should be slender to the hock, and from thence increase in thickness to the buttocks, which should be well developed. The carcass should be well rounded at each side, but level on the back and on the belly. There should be no hollows between the shoulder and the ribs, the line from the highest part of the shoulder to the insertion of the tail should be a perfect level. The flank should be full, the loins broad, and the tail finely formed and only partially covered with hair. The skin is a prime point : it must be covered with hair of a roan, or other *fashionable* color, and communicate to the hand of the experienced feeler, a peculiar sensation, which it is impossible to describe. With regard to this point, I cannot do better than quote the words of an experienced "handler" :—

"A nice or good judge of cattle or sheep, with a slight touch of the fingers upon the fatting points of the animal—viz., the hips, rump, ribs, flanks, breast, twist, shoulder score, &c. will know immediately whether it will make fat or not, and in which part it will be the fattest. I have often wished to convey in language that idea or sensation we acquire by the touch or feel of our fingers, which enables us to form a judgment when we are handling an animal intended to be fatted, but I have as often found myself unequal to that wish. It is very easy to know where an animal is fattest which is already made fat, because we can evidently feel a substance or quantity of fat—all those parts which are denominated the fatting points; but the difficulty is to explain how we know or distinguish animals, in a lean state, which will make fat and which will not—or rather, which will make fat in such points or parts, and not in others—which a person of judgment (*in practice*) can tell, as it were, instantaneously. I say *in practice*, because I believe that the best judges *out of practice* are not able to judge with precision—at least, I am not. We say this beast *touches* nicely upon its ribs, hips, &c., &c., because we find a mellow, pleasant feel on those parts; but we do not say soft, because there are some of this same sort of animals which have a soft, loose handle, of which we do not approve, because, though soft and loose, have not the mellow feel above mentioned. For though they both handle soft and loose, yet we know that the one will make fat and the other will not; and in this lies the difficulty of the explanation. We clearly find a particular kindliness or pleasantness in the feel of the one much superior to the other, by which we immediately conclude that this will make fat, and the other not so fat; and in this a person of judgment, and *in practice*, is very seldom mistaken."

In many respects the good points in a Shorthorn cow resemble those in the male of that breed, but in others there is considerable difference. As I have described in prose the excellencies which a bull should possess, I will now give a poetical summary of the good points of a cow of that breed, extracted from the *Journal of Agriculture*, and composed evidently by an excellent breeder and poet, Mr. Carr:—

The following features constitute, I trow,
The beau ideal of a short-horn cow :—
Frame massive, round, deep-barrell'd, and straight-back'd ;
Hind quarters level, lengthy, and well pack'd ;
Thighs wide, flesh'd inwards, plumb almost to hock ;
Twist deep, conjoining thighs in one square block ;
Loin broad and flat, thick flesh'd, and free from dip ;
Back ribs "well home," arch'd even with the hip ;
Hips flush with back, soft-cushion'd, not too wide ;
Flanks full and deep, well forward on the side ;
Fore ribs well-flesh'd, and rounded like a drum ;
Fore flanks that even with the elbow come ;
Crop "barrell'd" flush with shoulders and with side ;
Girth large and round—not deep alone, but wide ;
Shoulders sloped back, thick cover'd wide at chine ;
Points snug, well-flesh'd, to dew-lap tapering fine ;
Neck vein fill'd up to well-clothed shoulder-point ;
Arm full above, turn'd in at elbow-joint ;
Legs short and straight, fine boned 'neath hock and knee ;
Belly cylindrical, from drooping free ;
Chest wide between the legs, with downward sweep ;
Brisket round, massive, prominent, and deep ;
Neck fine at head, fast thickening towards its base ;
Head small, scope wide, fine muzzle and dish'd face ;
Eyes prominent and bright, yet soft and mild ;
Horns waxy, clear, of medium size, unfiled ;
Tail fine, neat hung, rectangular with back ;
Hide soft, substantial, yielding, but not slack ;
Hair furry, fine, thick set, of colour smart ;
Udder well forward, with teats wide apart.
These points proportion'd well delight the eye
Of grazier, dairyman, and passer-by ;
And these to more fastidious minds convey
Appearance stylish, feminine, and gay.

Breeds of the Ox.—The Shorthorned cattle are now generally regarded as the most valuable breed in these countries. They are the descendants of a short-horned breed of cattle which existed for centuries in the north-east of England. They were not held in much estimation, their flesh being coarse ; but the cows of this breed yielded abundance of milk. In the eighteenth century this breed, it is said, was greatly improved by a large infusion of blood from Dutch Shorthorns: but it is very doubtful that any such event took place, for during that period the importation of cattle into Great Britain was prohibited by very stringent laws. The present race of Shorthorns owe most of their valuable qualities to the brothers, Charles and Robert Colling, of the county of Durham. The former was the more successful breeder, and established the celebrated breed of Ketton Shorthorns. His whole process appears to have consisted in the careful selection of parents, and in "close" breeding. He must, however, have been an admirable judge of the good points of the ox, for beginning with animals not worth more on an average than £10 each, he produced in less than a quarter of a century a stock worth on the average £150 each. The most famous bull of Charles Colling's was Comet. The sale of this animal realised the handsome sum of 1,000 guineas. The bull Hubback is said by many writers to have been the great improver of Shorthorn blood. He was bought by Robert Colling for the trifling sum of £8; but although this animal was kept by both Collings for three years, there is good reason to believe that they made but little use of him. It would appear, indeed, that to the cows first used by the Collings—Lady Maynard, and young Strawberry—many of the good qualities of this breed are traceable. Shorthorns are now to be found in almost every part of the United Kingdom, capable of maintaining heavy stock. In Ireland the breed has been greatly improved, and it is gradually supplanting most of the other varieties.

Shorthorn males have a short, wide head, covered very often with short curly hair; the muzzle is taper; the ear rather long and narrow; the eye large, and bright, and mild. The shape is symmetrical, the carcass deep, the back level, ribs spreading out widely, and the limbs fine. The color is a mixture of red and white, sometimes a rich roan. The females are not so large in the head, which tapers more, and the neck is much thinner.

THE BREEDS OF STOCK. 67

The DEVONS are not so large as the Shorthorns. Their shape is symmetrical; fine head, horns of medium size, often tapering gracefully; rich red or orange red color; forequarters rather oblique. The meat of this breed is much esteemed: they yield excellent milk, but in rather limited quantity; and the bullocks answer the plough much better than many other kinds do. These animals arrive early at maturity.

The HEREFORDS are a rather small-boned breed; their horns are medium sized, straight or slightly curved upwards; their color is dark red; neat shoulders, thin thighs, and wide sirloin. They fatten well, but are not generally kept on dairy farms. In many respects they resemble the Devons.

The AYRSHIRES have a tapering head, fine neck, and large, bony, but not coarse carcass; flat ribs; short and rather ugly horns; their skin is soft, and covered with hair, which is usually red and white in spots. The Ayrshire cows are invaluable for dairy purposes.

The POLLED ANGUS, POLLED ABERDEENS, and POLLED GALLOWAYS are very large cattle, with big heads, unfurnished with horns. Their color is in general a decided black, but occasionally it exhibits a mixture of black and white. Their flesh is in general not of the best quality, but some of their crosses with Shorthorns yield excellent meat, and at an early age, too.

The KYLOES are a breed peculiar to the Highlands of Scotland. They are rather rough, but very picturesque animals, covered with long, shaggy hair. Their horns are rather long, and curve upwards. Their hair is differently colored—red, yellow, dun, and black, the latter being the prevailing hue. No variety of the ox yields a sweeter meat than the Kyloes, and other mountain breeds of these countries. The animals, however, arrive slowly to maturity, and in this respect there is great room for improvement. These mountain-bred animals are now transferred in large numbers to lowland tillage farms, where the fattening process is more expeditiously performed. There are excellent crosses between Shorthorn bulls and Highland cows.

LONGHORNED CATTLE are rapidly advancing towards extinction. At one time they were the chief breed kept by most farmers. In general they may be regarded as an inferior variety, being slow feeders, and producing rather coarse beef. They are, however, capable of great improvement, as instanced in the case of Bakewell's celebrated Longhorn herds.

The KERRYS are a diminutive breed, peculiar to Ireland. They have small heads, fine necks, fine horns of medium length, and curved upwards near their summits. They have a soft skin; the hair is generally black, interspersed with a few white streaks; sometimes their color is red, and occasionally brown. They are a very hardy race, being indigenous to mountains. Their flesh is very good, more especially if the animals have been kept on fattening food. The Kerrys are good milch cows.

The ALDERNEYS are a small race of oxen with deer-like faces. They exhibit various shades of red, white, brown, and roan. No cows yield better milk, or larger quantities of that fluid.

Sheep.—The different breeds of sheep are classified under three heads—viz., *Long-woolled, Short-woolled,* and *Middle-woolled.*

The LEICESTER is, perhaps, the most celebrated breed of sheep reared in these countries. It was immensely improved by Bakewell about a century ago, and the breed is often termed the Dishley, after the name of Bakewell's residence. This sheep has a wide, clean head, broad forehead, fine eyes, long, thin ears, thick neck, round body, deep chest, straight, broad back, high ribs, and muscular thighs. The wool is long, very thick, and fine. At from fifteen to eighteen months old, the Leicester weighs from 25 to 30 lbs. per quarter; but a fat animal often weighs from 38 to 40 lbs. per quarter. The fleece weighs from 6 to 8 lbs. This breed is well adapted for Ireland. It is reared on very poor land: but in order to maintain its good quality, this sheep requires abundance of food, and also good shelter during the winter.

The LINCOLN is distinguished for its large bones and strong muscles. Originally a gaunt and ugly animal, it has of late years been much improved. Indeed, the prices lately realised by Lincoln sheep are extremely high. The Lincoln has a long, white face, long body, and thick legs. The wool is long, thick, and moderately fine. The flesh of the Lincoln is lean, owing to its great muscular development. At fifteen months old it yields about 30 lbs. weight per quarter. It is said that a Lincoln wether has attained the weight of $304\frac{1}{2}$ lbs. The average weight of the wool of a hogget is $9\frac{1}{2}$ lbs.

The COTSWOLD breed arose in the Cotswold hills, in Gloucestershire. In this variety the skeleton is large, the chest capacious, the back broad and straight, and the ribs well arched. It has good quarters, and a finely-arched neck. It is distinguished by a large tuft of wool—"fore-top," on the forehead. It fattens early, and produces about 25 lbs. per quarter when fifteen months old, and 40 lbs. when two years old. The wool is rather coarse; its yield is about 8 lbs.

The CHEVIOT has a long body, long face, long legs, and long ears. The chest projects slightly, and is rather narrow. The forehead is bare of wool; the legs and face are white, sometimes approaching to a dun shade. Weight from 70 to 80 lbs.; weight of fleece, from 3 to 4 lbs. The wool is of excellent quality, and is used largely in the manufacture of tweeds. The Cheviot is a mountain sheep, and, as might be expected, its flesh is well flavored. There are several crosses of the Cheviot with the Leicester, the Southdown, and the Shropshire.

The SOUTHDOWN is generally regarded as the best breed for wool reared in these countries. It is indigenous to the chalk hills of Kent, Sussex, Hampshire, and Dorsetshire. It has a small head; its back is broad and straight; the ribs spring out at nearly right angles from the vertebræ. It is rather light in the fore-quarters, and full in the hind quarters.

Its chest is pretty deep; its face and legs are grey or brown. The wool of the Southdown is short, and extremely fine; the fleece weighs about 3 lbs. This sheep arrives early at maturity. It weighs at 15 months old about 80 lbs. The flesh is very well flavored.

THE SHROPSHIRE is said to combine in itself the good qualities of the Southdown, the Cotswold, and the Leicester. It resembles the Southdown more than any other breed, having the same grey, or brownish grey hue, and a similar shape. It is, however, larger than the Southdown, and yields a larger quantity of wool. This breed is becoming a great favorite in both England and Ireland.

The BLACK-FACED sheep is peculiar to Scotland. It is equipped with horns, has a bold long face, and possesses a tuft of wool on its forehead; its limbs are strong, and its body is somewhat long. The wool of this breed is very coarse, the fleece weighs about $3\frac{1}{2}$ lbs. The average weight of this sheep is 75 lbs., the quality of the mutton is excellent, but it is long before it becomes matured. There are several other breeds of the sheep, but they are of far less importance than those which I have described.

Breeds of the Pig.—There are several breeds of this useful animal, of which those known as BERKSHIRE and YORKSHIRE appear to be the greatest favorites. The Berkshire is black or dusky brown, very rarely reddish brown. It has a very small head. Its sides are extremely deep, and its legs very short. There are several sub-varieties of the Yorkshire. This breed is white, has a compact body, and very broad sides. The head is very small, somewhat like that of the Berkshire. Both Berkshire and Yorkshire pigs attain to the enormous weight of 1,000 lbs. The old Irish "racer" pig is the least profitable kind to keep, but fortunately it is, as a pure breed, nearly extinct.

Breeds of the Horse.—There are a great many breeds of horses. The Shetland pony is so small, that many specimens are no larger than a Newfoundland dog; on the other hand, Clydesdale horses sometimes attain to almost elephantine proportions. There is a wide difference between the bull-like Suffolk Punch and the greyhound-like *racer*. The English and Irish racer is said to owe its origin to a cross between the old English light-legged breed and the Arabian. The most valuable kind of carriage horse is the joint product of the draught-horse and the racer. The dray-horse of these countries has a large share of Flemish blood in him. The best horses for agricultural purposes are unquestionably the CLYDESDALE and the SUFFOLK PUNCH. The latter is perhaps to be preferred in most instances, especially on light lands. Very light and feeble horses are the most expensive variety on almost any kind of farm; for whilst they consume nearly as much food as the most powerful animals, and are therefore nearly as costly, they are incapable of effectively performing their work. A large proportion of the farm horses used by the small farmers of Ireland are totally unsuited for tillage purposes. On the other hand, there is no need to employ horses equal in size to the ponderous creatures that draw brewers' carts. Moderate sized horses, with well rounded, compact bodies, and muscular but not too heavy limbs, are the kind best adapted for farm purposes. In Ireland, where there are not fewer than 600,000 horses, a considerable infusion of blood from Clydesdales and Suffolk Punches is much required.

Hunters and Racers.—There is a strong tendency in the human mind to look with a regretful feeling to the past, and to compare it to the disadvantage of the present. It is a general belief with most people that the old time was the best time; that the seasons were more genial formerly; that provisions were cheaper and more abundant; that men were taller, and stouter, and healthier; that, in a word, everything was better in the days of yore than it is now, and that degeneracy and effeteness are the prevailing characteristics of our age. Philosophers, statists, and political economists tell us that all this regret for the "good old time" is mis-spent sympathy; for that we are in every respect superior—in physique, health, morals, and wealth—to our ancestors. On the whole, I rather incline myself to this comfortable philosophy; but we must admit that we have not progressed in all things since the times of our fathers.

In a work entitled "A Comparative View of the Form and Character of the English Racer and Saddle Horse during the Last and Present Centuries," published by Hookham, of Old Bond Street, London, it is proved very clearly that the English race-horse has sadly degenerated. The author very properly traces the cause of its decay to the avarice of the turfites: they look upon the noble animal as a mere gambling machine; and they sacrifice all its other qualities to the excessive development of that one which is likely to put money in their pockets. Formerly, gentlemen kept horses for their own sakes—for their admiration and enjoyment of one of the most beautiful, docile, and useful of animals. They were incessant in their efforts to develop into perfection all the really valuable points in the animal; and the result was, that the English and Irish racer of the last century was unmatched for strength, speed, and endurance. Models of this splendid race of horses are seldom to be found at the present time; but there are, perhaps, sporting men living who saw them in the celebrated Mambrino, Sweet Briar, and Sweet William. Those horses possessed compact bodies, capacious lungs, strong loins, large joints, and enormous masses of muscular tissue on the shoulder-blades and arms. They were good weight-carrying hunters as well as racers, and they could carry eight stones over a six miles heat, or twelve stones over a four miles one. The Irish horses, at least, were capable of safely carrying thirteen stones over what would now be considered a very ugly ditch, and could get over a long steeplechase in a style which would astonish the owners of the modern "weeds." Since the distance to be traversed by competing horses has been reduced from the old-fashioned three heats of four miles each to a single run of a mile or two, and also since the weight imposed upon the animals has been reduced to six or seven stones, from ten to twelve, the anatomical structure of the race-horse has undergone a remarkable and serious alteration. The back has become very long, the sides flat, the loins weak, the limbs long and very thin; and this alteration in structure has been attended by weakness of constitution and a remarkable tendency to disease. The modern horse has attained to a remarkable degree of rapidity of locomotion, but it has been at the expense of its vigor, endurance, and health; it can run with great velocity for a short distance, but in a four-mile heat, and mounted by a man of average weight, a mediocre horse of the style of the middle of the last century would come to the post long before the winner of the last St. Leger.

The decay of the breed of horses in this country is a serious matter, and the attention of all who are interested in the preservation of this animal should be earnestly and promptly directed towards discovering the means of regeneration. My remarks are directed towards racers and hunters. The quality of speed which they possess has been developed to an extent which is incompatible with the development of equally essential properties. Encouragement should be given to the production of weight-carrying hunters; steeple-chasing should be restored to its old state, when only a powerful horse had a chance of success. The quality of speed should be promoted in the animal up to a certain point; but when the development of this attribute begins to cause a loss of strength and endurance, it is high time to check it. There are a few horses at present which are strong and moderately fast: why should not steeple-chasing be of the kind which would call this style of animal into competition? Only a "weed" can now enter with any probability of success at a race of this kind; and when he has won it, of what use is he as a good hunter? What we want are good, stout, healthy horses, capable of carrying, in good style, twelve stones weight over a rough country; and the object of steeple-chasing should be the production of such a race of horses.

PART III.
ON THE MANAGEMENT OF LIVE STOCK.

SECTION I.
THE OX.

Breeding Cows.—The period of gestation in the cow is about nine months. The earliest time at which it is at all safe to breed from these animals is when they are one year and eight months old. Shorthorns breed early, whilst the mountain varieties are seldom in calf before they are three years old. The practice of very early breeding, though approved of by some extensive rearers of stock, is not to be commended for sound physiological reasons. Cows calve at all times of the year; but the most favorable time is near the end of winter, or in early spring. The cows should at this time be in fair condition—neither too fat nor too lean. Parturition should take place in a roomy, covered place, provided with abundance of clean litter. If such a place be not available, a nice paddock close to the house must answer. After having given birth to the calf, the cow should receive an oatmeal drink, or some warm and nutritious mash, and afterwards be liberally fed. The cow is usually allowed to run dry four or five weeks before calving: this period should not be curtailed; on the contrary, it would be better to extend it to six weeks, so as not to allow her condition to become too poor.

The Wintering of Young Stock.—There are certain localities wherein the rearing of young stock is one of the easiest tasks which devolve upon the farmer. Well-drained and shady fields, yielding abundance of sound herbage, and through which streams of *pure* water unceasingly flow, are just the proper *locale* for economically feeding young animals. But there are districts in which those favorable conditions do not exist; yet they are not better adapted to other uses. It is only the feeders of young stock in wet, moory, sandy, or undrained, heavy soils who really have cause for anxiety and incessant watchfulness. In rearing a calf the great object is to cause a rapid and uninterrupted increase in the weight of its body. At first the food of the animal should be furnished solely from the maternal founts; but at an early stage of its existence—about the third or fourth week—other food may wholly, or in part, be substituted for the natural aliment. It is important that no great interval should elapse between the hours of feeding. The digestive apparatus of the young animal is small, and its powers of assimilation are very energetic. The food with which it is supplied should, therefore, be given in moderate quantities, and very frequently. This is, in fact, what takes place when the calf is allowed free access to its dam; for the instant it feels a desire for aliment, the supply is at once available. Of course, there may be objections to this plan on the score of economy; but as a general rule, too much liberality cannot be exercised in feeding growing animals; and there is nothing more certain than that the calf which is illiberally fed will never be developed into a valuable, matured animal. When carefully tended from their birth, comfortably housed in winter, and abundantly supplied with nutritious food, it is sometimes wonderful the rapid progress which young stock make. Mr. Wright mentions a remarkable case of early maturity, which occurred in his own herd. A young steer, one year old, exhibited all the development of an animal twice its age. This bullock had been suckled for three months, whereby it had not only kept its calf-flesh, but gained and retained a step in advance. Its weight when only a year old was no less than 50 stones; and as the price of beef at the time was 8s. 9d. per stone, live weight, the carcass of the animal was worth £21 17s. 6d. Mr.

Wright offers this fact as a suggestive one to "those farmers who think of bringing up their calves on old milk, or who would otherwise stint their growth."

Supposing, then, that we have young stock which had been liberally treated when in their "baby" state, how are we to most economically maintain them throughout the winter? In the first place, they should be kept in warm sheds, and well sheltered from both rain and wind. Some authorities contend that exercise is necessary to young stock, and deny that a proper development of the muscles (lean flesh) can take place if they are cooped up like fattening turkeys during the winter. There is some truth in this opinion; and if the animals be designed for breeding or dairy purposes, their freedom of motion should only be partially restrained. On the other hand, if they be intended for an early introduction to the shambles, the less exercise they get the greater will be the profit on their keep. I have known cases where animals were closely housed for seven months, and yet their health did not appear to suffer in the slightest degree. In fact, so predominant are the vegetative functions of the ruminants over their nervous attributes, that the only essential conditions of their existence are adequate supplies of good air and food. That the health of these animals does occasionally suffer when the motions of their bodies are reduced to a *minimum* is quite true; but in most of these instances the real cause is, not the want of exercise, but the want of pure air. The greatest care should, therefore, be taken in the ventilation of the places where stock, whether old or young, are kept; and no economy of space or heat will compensate for the want of wholesome air. Under the fallacious idea that exposure to cold renders young stock hardy, many farmers turn them out to eat straw in the open fields in frosty weather. Treatment of this kind, instead of being productive of good, almost invariably lays the foundation of disease, which will manifest itself at some stage of the animal's growth. There are a few favored localities, such as those to which I have already alluded, where yearlings may be occasionally allowed a turn through the fields in winter ; but on cold clays, wet moors, and sandy soils the young stock should never be permitted to leave their sheds or courts from the time they are housed till late in the spring.

Young stock are best fed on good meadow hay and turnips, with a moderate supplement of oil-cake; this, however, is expensive feeding in many farms, and a little filling-in may be done with cheaper or more easily obtainable stuffs. A mixture of cut chaff, with pulped mangels, is a good substitute for the more costly hay; and particularly in the case of animals intended for breeding or for the dairy. The roots should be pulped, and allowed to remain until, owing to a slight fermentation, they become warm. This change takes place in from twenty-four hours to sixty hours, according to the temperature ; but the fermentation should not be carried farther than the earliest stage. The heated pulp should then be thoroughly mixed with the chaff, and the compound, after an hour or two, will be ready for use. A little chopped hay—no matter if inferior or slightly mildewed—may be substituted for the chaff, and turnips employed instead of the mangels, but the latter are the more desirable roots.

Until lately, the use of oil-cake was confined to fattening animals, but latterly it is freely given to calves, even when they are only a month old; and there is no doubt but that it is a suitable and economical food for store stock. It is, however, sometimes given in excess: from half a pound to two and a half pounds daily will be sufficient for animals under one year; and this addition to their food will be found to exercise a beneficial influence on them when they are placed in stalls for finishing. The experience of several eminent breeders has proved that fattening beasts, which had in their youth a supply of oil-cake, or its equivalent, invariably store up a larger portion of their food than those which had been reared on hay and roots only.

Mr. George Stodart, of Cultercullen, an Aberdeenshire farmer, describes, in the *Irish Farmer's Gazette*, his method of rearing calves :—

I occupy (says Mr. Stodart) a farm of 380 acres. I usually rear twenty-four calves yearly, and buy in sixteen one-year-olds. I generally breed from cross cows (the same as mentioned above), served by a pure Shorthorn bull. When the calves are dropped I put two calves to suck one cow for six months. In autumn, spring calves are put into the house upon turnips and straw, with about 1 lb. of oil-cake per day to each, until they are put out to grass in spring following, at which time they are one year old. Then, of course, they have grass in summer, and at the approach of winter they are again housed upon turnips and straw, which bring them to be two years old in spring. Now they are sent out to the best grass, and again brought into the house at the beginning of September, and fed on turnips and straw until the end of November or middle of December, when they usually fetch from £25 to £32 a-head. This year (1864), however, they will average £32 a-head. Before selling I give each 3½ lbs. of oil-cake per day for six weeks, and during this time they have swede turnips; at other times yellow. We give as much turnips at all times as they can eat.

Mr. Bowick, in his excellent paper on the rearing of calves, published in the Journal of the Royal Agricultural Society, gives the following information on this subject :—

We consider it desirable to allow the calf to remain with its dam for the first three or four days after calving.

Not much trouble is generally experienced in getting it to take to the pail. We find it better to miss the evening's meal, and next morning a very little attention induces the majority of them to partake of what is set before them. At most the guidance of the fingers may be wanted for the first meal or two.

As regards the quantity of milk which is needful to keep a moderately bred Shorthorn calf in a thriving condition, we have found the following allowance to come pretty near the mark, although the appetite of calves varies, both in individuals and at different times with the same animal :—

1st week with the dam ; or 4 quarts per day, at two meals.
2nd to 4th week, 5 to 6 quarts per day, at two meals.
4th to 6th week, 6 to 7 quarts per day, at two meals.

And the quantity need not, during the ensuing six weeks (after which it is weaned), exceed a couple of gallons per day. This implies that the calf is fed upon new milk only, and that no other feeding liquids are employed. But, in addition to the above, the calf will, towards the fourth week, begin to eat a little green hay ; and in a week or two later, some sliced roots, or meal, or finely crushed cake, mixed with hay-chaff ; and, if really good, creditable beasts are wanted—such as will realise £25 a-head from the butcher when turned two and a half years old—a little cake or meal in their early days will be found a desirable investment. In fact, we doubt not but 1 lb. of cake per day to the calf will make as much flesh as triple the quantity of cake at any period of after life. As regards meal, if that is given with the chaff, we prefer oatmeal, or barley-meal, or wheaten flour, but not the meal of beans or pease. Others may see it differently, but we believe beans to be too heating for any class of young stock. For roots, the best we know of is the carrot, grated and mixed with the chaff, or sliced thin with a knife and given alone. It is also, of all roots, the one which we find them most fond of, and which they will most readily take to. As soon as they can eat them freely, an immediate reduction in the supply of milk may be made.

In most articles it holds good in the end that "the best is the cheapest." So with the rearing of calves; the best class of food, or that above referred to, is found to give the greatest ultimate satisfaction. But practically the question often is, how to rear good calves with comparatively little new milk, a condition which circumstances often render almost imperative; for where dairy produce, in any other form, is the chief object, the calves stand in a secondary position, and are treated accordingly. But let us ask whether you cannot rear good stock under such circumstances also? We believe that this may be, and often is done. We manage to turn out from twenty-five to thirty calves annually—such as will pass muster anywhere—and never use at any one time more than six gallons of new milk daily. For this purpose, as well as to obtain a regular supply of milk for other purposes, the calves are allowed to come at different periods, extending from October to May. Hence the calf-house has generally a succession of occupants throughout the season; and as one lot are ready to be removed, and placed loose in a small hovel, with yard attached, others fill their places. We begin with new milk from the pail, which is continued for a fortnight after leaving the cow. Then skim-milk —boiled, and allowed to cool to the natural warmth—is substituted to the extent of one-third of the allowance. In another week the new milk is reduced to half, and at the same time, not before, boiled linseed is added to the mess.* As soon as they take freely to this food, the new milk may be replaced with that from the dairy, and the calf is encouraged to indulge in a few sliced carrots and the other dry foods named.

Mr. Murray, of Overstone, thus states the expense of

---

* Five pounds of linseed will make about seven gallons of gruel, and suffice for five good-sized calves; considerable allowance must, however, be made for differences of quality in the linseed, that from India not being gelatinous enough, and therefore boiling hard, instead of "coming down kindly."

rearing the calf until it is two years old, when, after the weaning process is completed, it is turned out to grass :—

During the summer they have the run of a grass paddock during the day, but return regularly to their yards at night ; the following winter they are kept in larger yards, and which contain a greater number of animals. Their bill of fare for this winter is 2 lbs. of oil-cake, half a bushel of cut roots, with cut chaff *ad libitum*. The chaff has a small quantity of flour or pollard mixed with it, is moistened with water, and the whole mass turned over ; this is done the day previous to using it. By this means they eat the chaff with more relish, and moistening it prevents the flour being wasted. They are put to grass the following summer, generally from the 15th to the 20th of May, or as soon as the pastures are in a state to receive them ; they remain there on second-rate land till about the end of October, when they are brought home and tied up in the stalls. The daily allowance is then 4 lbs. linseed-cake, 4 lbs. flour—$\frac{3}{4}$ bean, $\frac{1}{4}$ barley—1 bushel of cut roots with cut chaff ; the flour and chaff is mixed as already described. At about the end of December the quantity of cake is increased to 8 lbs., and the flour to 6 lbs. ; this they continue to receive till they are sold to the butcher during the months of March and April, when they weigh, on an average, 90 stones of 8 lbs. per bullock, and under two years and six months old. At this season of the year beef generally makes 5s. per stone—we often make 9s.—but taking that as an average would make the value of each beast £22 10s. The cost of keeping to this age will be as follows :—

	£	s.	d.
One calf	2	0	0
Milk, &c., nine weeks	1	5	0
Cake, grass, &c., forty-three weeks, at 1s. 6d.	3	4	6
Second year, November till May, cake, flour, roots, &c., 2s. 6d. per week, for twenty-six weeks	3	5	0
May till November, grass, twenty-six weeks, at 2s. 6d.	3	5	0
Third year, November till April, twenty weeks, at 8s.	8	0	0
	£20	19	6

Which leaves a gain to each animal of £1 10s. 6d., besides the manure.

Shelter of Stock.—The great diminution of temperature, and the falling off in the supply of herbage, that are coincident with the close of the autumn, render it necessary to remove our cattle from the open fields, and provide them with some sort of shelter during the winter months and early part of the spring.

The particular period at which this change of quarters takes place of course varies, and is, in fact, altogether dependent upon the character of the season. There are some years in which there is, so to speak, a kind of relapse of the summer, November being bright and warm, instead of, as is usually the case, cold and foggy. In such a year there is some herbage to be picked up until the very end of December. On the other hand, the latter part of October is often very wet, and October frosts are by no means uncommon. Tempestuous, biting winds in November, or torrents of rain, or both, tell severely upon the poor animals in the fields, even where there is abundance of herbage; and hence, should such weather take place at the latter part of October, the true economy would be to remove the animals at once to sheltered places.

Nothing lowers the temperature of the surface so rapidly as a cold wind. Captain Parry, one of the explorers of the Arctic regions, states that his men, when well clothed, suffered no inconvenience on exposure to the low temperature of 55 degrees below zero, provided the air was perfectly calm; but the slightest breeze, when the air was at this temperature, caused the painful sensation produced by intense cold. I could adduce the experience of many practical men in favor of the plan of affording shelter to animals, but more especially to those kept in situations much exposed to winds. Mr. Nesbit relates a case bearing on this point :—A farmer in Dorsetshire put up twenty or thirty sheep, under the protection of a series of upright double hurdles lined with straw, having as a sort of roof, or lean-to, a single hurdle, also lined with straw. A like number of sheep, of the same weight, were fed in the open field, without shelter of any kind. Each set was fed with turnips *ad libitum*. The result was, that those without shelter increased in weight 1 lb. per week for each sheep, whilst those under shelter, although they consumed less food, increased respectively 3 lbs. per week.

As a general rule, the latter part of October, or early in November, is the time for the removal of live stock from the pastures to the shelter of the farmstead. In England and Scotland the transference is seldom delayed after these dates; but in Ireland it is no uncommon thing to see the animals grazing very much later in the year—a circumstance which the lateness and mildness of our climate account for. But whatever the date may be, the importance of such shelter is universally recognised, even by those who most neglect it and are least acquainted with the principles upon which its necessity depends. The more important of these principles have already been explained, but they may be here summarised as follows :—

1. A certain amount of warmth is an indispensable condition for the maintenance of the life of animals.

2. The internal heat of the bodies of animals is supplied by the chemical combination which takes place between the oxygen of the atmospheric air which they inspire and certain of the constituents (carbon and hydrogen) of the food which they consume, or, to speak more accurately, of the tissues of their bodies, which are formed out of their food. It is very much in the same way in which our houses are heated by the burning of coal, turf, or wood in their fire-places, since the heat derived in the latter case is obtained from a similar source as in the former one—namely, by the union of the oxygen of the air with the carbon and hydrogen of the fuel. The only real difference between the two kinds of combustion is, that in respiration the process is conducted with an extreme degree of slowness, whilst in the ordinary fire the combinations take place rapidly, and the heat being evolved in a much shorter time is proportionately the more intense.

3. The temperature of the external parts of the animal body varies with the nature and quantity of the food supplied to it, and also depends upon the state of the weather and the character of the protection afforded to it.

The colder the air, the greater will be the quantity of food required, and the more complete the shelter. In other words, a diminution of temperature, no matter how caused, will necessitate an increased amount of food and more perfect shelter, in order to maintain at the proper degree of heat the fluids of the body. It is only the external parts of the body that become cold : so long as the animal is in health its blood always maintains the same degree of temperature ; but in cold weather the blood is subjected to a greater cooling power than it is in warm weather, and this cooling power it can only resist by taxing more extensively the heat-producing resources of the body.

4. Exposure to wet, even in warm weather, will tend to reduce the temperature of the body, since the conversion of water into vapor can only be effected at the expense of heat, which heat must be in great part extracted from the body of the animal itself.

5. No possible increase of food, however nutritious it may be, can suffice to keep up the due warmth and healthy condition of the animal frame in winter, if shelter from cold and rain be not simultaneously effected. On the contrary, an animal well protected from the winter blasts will require much less food than if it were placed in an exposed position. The reason of this is, that the amount of food which an animal exposed to great cold consumes to maintain the temperature of its body would, under opposite conditions, be stored up in the form of permanent "increase"—beef or mutton for the butcher, in fact.

The fat-forming constituents of the food of stock are in no case converted into permanent fat, except when they exceed in quantity the amount required to keep up the internal heat of the animal; but when this is constantly reduced by exposure to a wintry temperature, the food becomes insufficient for even that purpose, no matter how much aliment is given. What, then, must not be the condition of the unfortunate animals whose fate it is to be the property of a farmer who neither shelters them from the weather nor provides them with a sufficient quantity of nourishing food !

Milch Cows.—When dairy-farming is conducted on pure pastures, the cows are altogether dependent upon the grasses; and in winter, the animals suffer much from scarcity of food. This is the very worst system of cow-keeping, but it is prevalent amongst many small farmers in Ireland, and is to be met with even in England and Scotland. I am strongly of opinion that it would be far more economical to keep cows (and other cattle) altogether in the house, and feed them with cut grass, than to allow them to remain out altogether in the field. There are several disadvantages resulting from the depasturing of cows. In the warm weather, the animals are greatly annoyed by the attacks of flies : there is a considerable waste of muscle, caused by the movements of the animals whilst in search of their food ; and the excrements of the animals and their footmarks injure a large portion of the grass. It may be somewhat troublesome and expensive to cut the grass, and convey it from the field to the house ; but the labor and the cost will be more than repaid by the greatly-increased yield of food. A grass-field, mowed, will produce from 20 to 30 per cent. more food than it would if it were trampled upon and soiled by cattle. Exercise for an hour or two in the cool of the evening, or early in the morning (during the hot weather), will be quite sufficient to keep the animals in health. This may be taken in a field, better in a paddock, best of all in a roomy yard. When cattle are supplied with cut grass, or clover, care should be taken not to give it to them when very wet, for otherwise there is danger of the excessively moist herbage producing the *hoove.* Neither should large quantities of the green food be given to them—the supply should be " little and often." Should the food be too succulent, the addition of a little straw will correct its laxative effects. When the stock is about passing from the winter keep to summer food, the transition should be gradual; a well-made compound of straw or hay with grass (natural or artificial) is much relished by cows. A supply of good water is absolutely necessary; but sufficient attention to this important point is seldom given. Cooked food is well adapted for milch cows. Mangels, kohl-rabi, and cabbages are each of them better food than turnips, as the latter is apt to impart a disagreeable flavour to the butter. Three feeds in the day is a sufficient number for cows. The first meal should be early in the morning, and may consist of roots, mixed with straw or hay. Some feeders prefer using dry fodder, or cooked food of some kind, and not raw roots. The second meal is given at mid-day, and the third in the evening. The daily allowance of roots varies from 2 to 8 stones, depending upon the quantities of other foods used. Mr. Horsfall's diet is as follows :—Hay, 9 lbs.; rape-cake, 6 lbs.; malt-combs, 1 lb.; bran, 1 lb.; roots, 28 lbs. These substances are mixed and cooked, and the animals receive them in a warm state. In addition to this food, Mr. Horsfall's cows get bean-meal—a cow in full milk 2 lbs., others from ½ lb. to 1½ lbs.; cost per week per cow, 8s. 7d.* Mr. Alcock, of Skipton, feeds his cows as follows :—Raw mangels, 20 lbs.; carob beans, 3 lbs.; bran and malt-combs, 1¾ lbs.; bean-meal, 3½ lbs.; rape-cake, 3 lbs.; per diem. A steamed mixture of wheat and bean straws and shells of oats *ad libitum*. Oats, to the extent of 2 or 3 lbs. daily, are an excellent food for cows.

An important point in dairy economics is the feeding of the cows at *regular* intervals. If the usual time for the feed be allowed to pass, the animals are almost certain to become very uneasy—to *worry;* and every feeder knows, or ought to know, that a fretting beast will neither fatten nor yield milk satisfactorily. The cow-house ought to be kept as clean as

* " Journal of the Royal Agricultural Society," vol. xxxix.

possible; and the excreta, therefore, should be removed several times a day.

Mr. Harvey, of Glasgow, has probably one of the largest dairies in the world. His cow byres, 56 yards long, and from 12 to 24 feet wide—according as one or two rows of cows are to be accommodated—stand closely packed, the whole surface of the ground being thus covered by a kind of roof. From 900 to 1,000 cows are constantly in milk. They are fed during winter partly on steamed turnips (7 tons being steamed daily in order to give one meal daily to 900 cows), partly on coarse hay, of which, as of straw, they get between 20 and 30 lbs. a day each. They are also fed on draff, of which they receive half a bushel daily each; on Indian-corn meal, of which they have 3 lbs. daily each; and on pot-ale, of which they receive three times a day nearly as much as they will consume, *i.e.*, from 6 to 10 gallons daily. During the summer they are let out, a byreful at a time, for half a day to grass, and on coming in receive their spent malt and still liquor, and hay in addition. They are managed, cleaned, and fed by two men to each byre holding about 100 cows. The milking is done three times a day, by women who take charge of 13 cows in full milk, or double that number in half milk, apiece. Between 4 and 5 o'clock a.m. (taking the winter management), the byres are cleaned out, and the cows receive a "big shovelful" of draff apiece, and half their steamed turnips and meal, and a "half stoupful," (probably 2 gallons) of pot-ale. They are milked very early. At 7 they receive their fodder-straw or hay. At 10 they get a "full stoupful" (probably 3 or 4 gallons) of pot-ale. They are milked at noon. At 2 p.m., or thereabouts, they are foddered again, and at 4 p.m. receive the same food as at the morning meal. They are again milked at 5 to 6, cleaned out and left till morning. The average produce is stated to be 2 gallons a day per cow.

Mrs. Scott, of Weekston, Peebles, who keeps one of the best managed dairy farms in the United Kingdom, thus conducts her operations in the winter :—At 6 o'clock in the morning the cows are well wiped or scrubbed, have their bedding removed, and receive each about 4 or 5 lbs. of straw. At 8 o'clock the cows are milked, and Mrs. Scott examines each to ascertain whether or not the milk-maid has left any fluid in the udder—and woe betide the careless maid if her work has been carelessly done! At 10 o'clock a barrowful of turnips is divided amongst three cows, and when these roots are not available, a quantity of peas or bean meal, with a pint of cold water, takes their place. At 1 o'clock the cows are allowed out to be watered, and during their absence from the byre it is thoroughly cleansed and ventilated. When the state of the weather prevents the cows from being turned out, they receive twice a day a handful of oatmeal diffused throughout three pints of water—a handful of salt being given in the first of these drinks. When the cows return to the byre, they receive each about 4 or 5 lbs. of straw, and at 4 or 5 o'clock an evening meal of turnips equal to their morning feed. At 8 o'clock a "windling" of meadow hay is given to each pair of cows, the quantity being always regulated according to the requirements of each cow. The cows upon calving receive, in addition to this allowance of hay, half a pailful of boiled turnips, mixed with a quart of peas or bean-meal. This mess is given in a lukewarm state. Mrs. Scott's system may be thus epitomised: Regularity in feeding; sufficient but not excessive food; regularity in milking; and minute attention to cleanliness and ventilation.

Stall-feeding.—What becomes of the 90 per cent. of the weight of the non-nitrogenous constituents of the food of the sheep, and of the 80 per cent. of that of the nutriment of the pig, which they consume but do not store up? I have already partly answered this question. This portion of the food is chiefly expended in the production of the heat with which the high temperature of the animal's body is maintained. Part of it, no doubt, passes unchanged through its body, either owing to its indigestibility, or to its being given in excess. The quantity of non-nitrogenous matters consumed by a man is influenced greatly by the temperature of the air which he habitually breathes, and by the nature of the artificial covering of his body; there may be other conditions at present unknown to us, but these are amongst the chief ones. Now, as there is sufficient reason to lead us to believe that the consumption of carbonaceous food by the lower animals is influenced in the same way by the temperature of the medium in which they exist, the question naturally suggests itself, would it not be cheaper to maintain the heat of the animal by burning the carbon of cheap coal or turf outside its body, than by consuming the carbon of costly fat within it? The answer to this question is not so simple as at first sight it appears to be. We must not consider that, because 10 lbs. weight of carbon, as coal, costs but a penny, whilst an equal weight of the same element in starch costs twenty pence, heat may be furnished to a fattening animal twenty times cheaper by the combustion of coal than by that of starch. No doubt the amount of heat evolved by the conversion of a pound-weight of carbon into carbonic acid is the same, whether it be a constituent of starch or of coal; but the application of the heat so produced is less under our control in the latter case. All the heat evolved during the combustion of the starch within the animal's body is made use of; whilst a very large proportion of that developed by the combustion of coal in a furnace cannot in practice be applied to the purpose of heating the animal's body.

It is only the handiwork of the Creator which is perfect, and no machine constructed by the skill of man, for the direction of force, can rival that wondrous heat-producing, force-directing mechanism — the animal organism. According to Dumas, the combustion of about $2\frac{1}{2}$ lbs. of carbon in a steam-engine is required to generate sufficient force to convey a man from the level of the sea to the summit of Mont Blanc; but a man will ascend the mountain in two days, and burn in his mechanism only half a pound of carbon. There is no machine in which heat and force are more completely made available than the animal organism; and were it not—thanks to the influence of antediluvian sunshine—that the carbon of fuel in these countries is so very much cheaper than the carbon of food, there is no doubt but that the cheapest mode of keeping an animal warm would be to allow it to burn its carbon within its body. As the matter stands, however, there is no question as to the advisability of keeping fattening animals in a warm place. If the temperature of the stall be equal to that of the animal's body there will be less food consumed in the increase of its fat; because less of the fat-forming materials will be expended in the production of heat. In this sense, therefore, heat is an equivalent to food, but only within certain limits; because heat is developed in large quantity within the animal body independently of the temperature of the air. There is, therefore, no object to be attained by having the stalls heated beyond 70 or 80 degrees. Indeed, it is to be questioned whether or not stalls artificially heated are ever properly ventilated. If they be not, the health of the animal will suffer, and its appetite—so essential a point in fattening stock—will become impaired. We may conclude—firstly, that animals, when fattening, should be kept at a temperature not under 70 degrees nor above 90 degrees Fahrenheit; secondly, that the mode of heating must be such that there is as little wasteful combustion of fuel as is possible under the circumstances; and, lastly, that no motives of economy of fuel should prevent the feeding places from being thoroughly ventilated.

Stall-feeding is not so extensively carried on in Ireland as it is in Great Britain. There is a general impression that it does not pay in the former country; but if such be the case, it is simply owing to the want of skill on the part of the Irish feeders.

The cattle intended for stall-feeding should be removed (if out) from the field in October, and put into the house, or court, or crib, or hammel, as the case may be. They are fed upon roots, straw, hay, grain, and artificial food. The greatest skill is required in their treatment. It is a nice point to determine which foods are the most economical, and also to ascertain in what foods excessive proportions of certain nutritive elements exist. Sufficient food should be given; but any approach to waste should be avoided. Three feeds a day are usually given, and should be supplied at the same hours each day. For about two weeks the animals are furnished with white turnips *ad libitum;* but after the expiration of that time they receive Swedish turnips, straw, and grain, or oil-cake. Late in the season mangels will replace turnips. Almost every extensive feeder now uses oil-cakes in large quantities; but when oats are low in price, they will in general be found a cheap equivalent for a large proportion of the oil-cake. Different feeders have different dietaries, and the nature of the aliments supplied to fattening stock depends very much upon the market prices of food-stuffs, and the locality in which the feeding-house is situated. The following dietaries are but examples of the methods of feeding adopted in different districts and by different persons :—

Mr. McCombie, of Tillyfour, fattens from 300 to 400 beasts annually, and obtained for them in 1861 £35 per head. He never exceeds 4 lbs. of oil-cake per diem, nor 2 lbs. of bruised oats, for each beast. He gives as much turnip and straw as they can consume. He realises £12 per acre in feeding on Aberdeen and Swedish turnips.

"For fatting cattle," says Mr. Edmonds, of Cirencester, "I should recommend two parts hay and one part straw, or in forward animals three parts hay and one part straw cut in chaff. Those of average size will eat somewhere about five bushels per day, with 4 lbs. to 5 lbs. oil-cake, and half a peck of mixed meal, barley and peas, or beans, and, if cheap, a proportion of wheat also, to be increased to one peck per day in a month or six weeks after they have come to stall, the oil-cake and meal to be boiled in water for half-an-hour or three-quarters, and thrown in the form of rich soup over the chaff, and well mixed, to which add a little salt."

Colonel M'Douall, of Logan, Wigtonshire, gives 3 lbs. of

bean-meal and 3 lbs. of cut straw cooked together, and 84 lbs. of Swedish turnips.

According to the researches of Messrs. Lawes and Gilbert, an ox weighing 1,400 lbs. ought to gain 20 lbs. weekly when fed under cover with 8 lbs. of crushed oil-cake, 13 lbs. of chopped clover hay, and 47 lbs. of turnips. The chemical constituents (in a dried state) of this allowance are as follows:—

	Ounces.
Fat-formers, or heat givers	232
Flesh-formers	55
Mineral matter	29

Cost of Maintaining Animals.—The animal mechanism, which exhibits the least tendency to fatten, is the most costly to keep in repair, in relation to the work performed by it. If, for example, a sheep store up in its increase one-fifth of its food, then the remaining four-fifths are expended in preserving it alive, and their cost represents, so to speak, the expense of preserving the animal's body in repair. If another sheep store up only one-tenth of its food, then the cost of its maintenance may be said to be double that of the animal which retains the larger proportion of its nutriment in the form of flesh. Of course in both cases the value of the manure will to a great extent compensate for the cost of the food expended in merely keeping the animal alive; but that does not affect the proposition, that the less food expended by an animal in carrying on its vital functions the more valuable is it as a "meat-manufacturing machine." From the moment it is brought into the world until it is "ripe" for the shambles, an animal should steadily increase in weight: every week that it does not store up a portion of its food in permanent increase is the loss of a week's food to the feeder; for all the fodder consumed during that time by the animal is, so to speak, devoted to its own private purposes. Sheep overcrowded on pastures, milch cows on "short commons," calves kept on bulky innutritious food, are all so many sources of positive loss to the feeder—and as many proofs that he who aspires to be a successful producer of meat, must, in one respect at least, be a devout believer in the doctrine of Progressive Development.

Cooking and Bruising Food.—The cooking, or the otherwise preparing, of the food of the domesticated animals is a subject which until recently was completely ignored by the vast majority of stock feeders. It is now, however, beginning to attract a fair amount of attention; and no doubt ere long the best modes of treating the food of cattle will be discovered.

As might be expected from our limited experience of the subject, there exists considerable difference of opinion relative to the proper method of cooking cattle food; and there are many very extensive feeders who object to the plan altogether, and contend that as the food of the inferior animals is naturally supplied to them in a raw condition, it would be quite unnatural to give it to them in a cooked state.

Whatever difference of opinion there may be with regard to the propriety of cooking the food of stock, we believe there ought not to be a doubt as to the desirability of mechanically treating the harder kinds of feeding stuff. It is quite evident that a horse fed upon hard grains of oats and wiry fibres of uncut hay or straw must expend no inconsiderable proportion of his motive power in the process of mastication. After a hard day's work of eight or ten hours he has before him the laborious task of reducing to a pulp from 12 lbs. to 20 lbs. weight of exceedingly hard and tough vegetable matter; and as this operation is carried on during the hours which should be devoted to rest, the repose of the animal is to some extent interfered with. Indeed, it not unfrequently happens that a horse, after a hard day's work, is too tired to chew his food properly; he consequently bolts his oats, a large proportion of which, as a matter of course, passes unchanged through the animal's body.

In order to render fully effective the motive power of the horse, it is absolutely necessary to pay attention to the condition, as well as to the quantity and quality of his nutriment. The force wasted by a horse in the comminution of his food, when composed of whole oats and uncut hay and straw, cannot, at the lowest estimate, be less than that which he expends in an hour of ordinary work, such as, for example, in ploughing. The preparation of his food by means of water or steam power, or even by animal motive power, would economise by at least 50 per cent. the labor expended in its mastication; and this would be equivalent to nearly half a day's work in each week, and, consequently, a clear gain of so much labor to the owner of the animal. In the present time of water-power and steam-power corn-mills, one man is able to grind the flour necessary for the support of several thousand men; in early ages the labor of one person in the grinding of wheat served but to supply the wants of twenty others. In both cases machinery was employed for reducing the grain to flour; but in the one case, the mechanisms employed were more than a hundred times more effective than in the other. But even the most imperfect flour mill is by far a more economical system of comminuting corn than the jaws of animals; and if every man were obliged, as the horse is, to grind his corn by means of his teeth alone, he would find his powers for the performance of other kinds of labor considerably lessened.

It has been urged as an objection to the use of bruised oats by horses, that they exercise in that state a laxative influence upon the animal's bowels. I doubt very much that such is frequently the case, when the animal is fed only upon oats and hay and straw; but even if the oats produce such an effect, the addition of a small proportion of beans—the binding properties of which are well known—will obviate the disadvantage.

The desirability of mechanically acting upon soft food is not so apparent as the necessity for the bruising of oats is. Roots are so easily masticable that if they are rendered more so there is danger of their being so hastily swallowed as to escape thorough insalivation, which is so necessary to ensure perfect digestion. To guard against this danger, perhaps the best way would be to give pulped mangels and turnips mixed with cut straw; a mixture which could not easily be bolted. Mr. Charles Lawrence, of Cirencester, who is a great advocate for the cooking of food, and has frequently published his experience of the benefits derivable therefrom, thus describes his method of combining pulped roots with dry fodder :—

We find that, taking a score of bullocks together fattening, they consume per head per diem three bushels of chaff, mixed with just half a cwt. of pulped roots, exclusive of cakes of corn ; that is to say, rather more than two bushels of chaff are mixed with the roots, and given at two feeds, morning and evening, and the remainder is given with the cake, &c., at the middle-day feed, thus:—We use the steaming apparatus of Stanley, of Peterborough, consisting of a boiler in the centre, in which the steam is generated, and which is connected by a pipe on the left hand with a large galvanised iron receptacle for steaming food for pigs, and on the right with a large wooden tub, lined with copper, in which the cake, mixed with water, is made into a thick soup. Adjoining this is a slate tank, of sufficient size to contain one feed for the entire lot of bullocks feeding. Into this tank is laid chaff with a three-grained fork, and pressed down firmly; and this process is repeated until the slate tank is full, when it is covered down for an hour or two before feeding time. The soup is then found entirely absorbed by the chaff, which has become softened and prepared for ready digestion.

Mr. Wright, near Dunbar, gives the following account of an experiment with pulped roots and straw and oil-cake. It appears to prove the superiority of mixed foods over the same foods consumed separately :—

Two lots of year-old cattle were fed ; the one in the usual way—sliced turnips and straw, *ad libitum*—the others with the minced turnips, mixed with cut straw. The first lot consumed daily 84 lbs. sliced turnips, 1 lb. oil-cake, 1 lb. rape-cake, ¼ lb. bean-meal, broken small and mixed with a little salt, and what straw they liked. The second lot ate, each, daily, 50 lbs. minced turnips, 1 lb. oil-cake, 1 lb. rape-cake, ¼ lb. bean-meal, and a little salt, the whole being mixed with double the bulk of cut straw or wheat chaff. In spring, the lot of cattle which had the mixed food were in good condition, and equally well grown as others, though they had consumed in five months two tons less of roots apiece. The reporter does not advise the mincing process to be commenced when cattle are very forward in condition, as any change of food requires a certain time to accustom the animals to it, and in the meantime fat cattle are apt to fall off in condition. It ought to be begun when they are young and lean.

Mr. Duckham, of Baysham Court, Ross, Herefordshire, says :—

The advantages of pulping roots for cattle are—1st, Economy of food; for the roots being pulped and mixed with the chaff, either from threshing or cut hay or straw, the whole is consumed without waste, the animals not being able to separate the chaff from the pulped roots, as is the case when the roots are merely sliced by the common cutter, neither do they waste the fodder as when given without being cut.

2. The use of ordinary hay or straw. After being mixed with the pulp for about twelve hours, fermentation commences, and this soon renders the most mouldy hay palatable, and animals eat with avidity that which they would otherwise reject. This fermentation softens the straw, makes it more palatable, and puts it in a state to assimilate more readily with the other food. In this respect I think the pulper of great value, particularly upon corn farms where large crops of straw are grown, and where there is a limited acreage of pasture, as by its use the pastures may be grazed, the expensive process of haymaking reduced, and, consequently, an increased number of cattle kept. I keep one-third more, giving the young stock a small quantity of oil-cake, which I mix with the chaff, &c.

3. Choking is utterly impossible, and I have only had one case of hoove in three years, and that occurred when the mixture had not fermented.

4. There is an advantage in mixing the meal with the chaff and pulped roots for fattening animals, as thereby they cannot separate it, and the moisture from the fermentation softens the meal and ensures its thorough digestion, whereas, when given in a dry state without any mixture, frequently a great portion passes away in the manure.

On the value of the process for a grazing farm with but a small quantity of plough-land, Mr. Corner, of Woodlands, Holford, Bridgewater, thus speaks :—

My plan is, first commencing with the grazing beasts, to cut about an equal quantity of hay and straw and mix with a sufficient quantity of roots (mostly mangel) to well moisten the chaff; and as the beasts advance in condition, I lessen the straw and increase the hay, and in their further progress I mix—in addition to all hay, chaff, and roots—from 6 to 10 lb. per day to each bullock of barley and bean-meal, according to its size—and I have them large sometimes. I sold last week for the London market a lot of Devon oxen of very prime quality, averaging in weight upwards of 100 stone imperial each.

For my horses, cows, yearlings, and oxen—the latter to be kept in a thriving condition, and turned to grass, and kept through the summer for Christmas, 1860—I cut nearly all straw, with a very small quantity of hay, and this the offal of the rick. These also have as many pulped roots as will moisten the chaff, except the horses, and to them I give, along with bruised oats, just enough roots to keep their bowels in a proper condition.

To the two or three-year-old beasts I give some long straw and a part chaff, and the offal (if any) of the food of the above lots of stock.

My farm is but a small one—under 200 acres. My predecessor always mowed nearly all the pastures for hay, which is about half the farm, and with this scarcely ever grazed any beasts, and kept but very few sheep. Since my occupation I scarcely ever exceed ten acres of meadow with one field of seeds for hay. I keep from 250 to 300 large-size Leicester sheep, and graze from 20 to 25 large-size beasts a year, with other breeding stock in proportion.

I consider the pulping of roots is better for fatting pigs than anything else. My plan is to have a large two-hogshead vat as near the pulping machine as possible, so as to fill it with a malt shovel as it comes from the machine; at the same time I keep a lad sprinkling meal (either barley or Indian corn) with the roots; and this is all done in fifteen or twenty minutes. It is then ready for use, to be carried to the pigs in the stalls alongside the fatting beasts. I never could fatten a pig with profit until I used pulped roots.

Although the practice of cooking food has been advocated by several eminent feeders, it has been condemned by others. Mr. Lawes is not favorable to the cooking of food unless when it is scarce. The results of Colonel M'Douall's experiments go to prove that cattle can be more economically kept upon a mixture of raw and cooked foods than upon either raw or cooked fodder given separately. One meal of cooked food and two feeds of raw turnips gave better results than three feeds of raw turnips; whilst two cooked feeds and a raw one resulted in a loss.

The fermentation of food, if not the best, is certainly the cheapest mode of preparing it. If the process be not pushed too far the loss of nutriment sustained is inconsiderable. When a mixture of straw and roots is fermented, the hard fibres of the latter are, to a great extent, broken up, and the nutrient particles which they envelop are fully exposed to the action of the solvent juices of the stomach.

A great advantage in cooking or fermenting food is that the most rubbishy materials can be used up. Indeed, as a general rule, the better soft food is, the less the necessity for cooking it; but washed out hay and hard, over-ripened straw are of but little value, except when cooked and given in combination with some agreeably-flavored substance.

VALUE FOR FEEDING PURPOSES OF VARIOUS FOODS.*

MATERIAL.	Cost. Per ton.			Cost. Per 100 lbs.		100 lbs. contain. Oil.	Starch, Sugar, &c.	Oil, Starch, &c., computed as Oil.	Nitrogen. Weight.	Nitrogen. Value.	Phosphoric Acid. Weight.	Phosphoric Acid. Value.	Potash. Weight.	Potash. Value.	Value of Nitrogen, Phosphoric Acid, and Potash.		Deduct Nitrogen for perspiration.	Net Value for Manure.	
	£	s.	d.	s.	d.	lbs.	lbs.	lbs.	lbs.	d.	lbs.	d.	lbs.	d.	s.	d.	d.	s	d.
Meadow-hay ...	4	0	0	3	7	2·68	39·75	24·63	1·48	10·62	0·90	1·35	1·50	4·50	1	4½	2 1/12	1	2¼
Wheat-straw ...	1	15	0	1	7	0·50	32·0	18·50	0·42	3·0	0·14	0·21	0·65	2·16	0	5	½	0	5
Swedish Turnips	4	10	0	4	0	2·0	60·0	35·0	2·40	17·28	0·80	1·20	2·25	6·75	2	1¼	3½	1	9¾
Oil-cake	9	6	8	8	4	12·0	38·0	33·0	5·0	36·0	2·25	3·37	1·75	5·25	3	8½	7¼	3	1¾
Beans	9	6	8	8	4	2·0	42·0	25·30	4·45	32·0	0·86	1·29	1·11	3·33	3	0½	6½	2	6
Indian Meal ...	9	6	8	8	4	7·0	60·0	40·0	2·25	16·20	0·19	0·28	0·17	0·51	1	5	3¼	1	1¾
Carob, or Locust Bean...........	9	6	8	8	4	6·76	57·0	35·0	0·64	3·75	No analysis of ash.				say	5¾	—	0	5

* From Mr. Horsfall's Essay on Dairy Management, in "Journal of Royal Agricultural Society," vol. xviii., part i.

98   THE MANAGEMENT OF LIVE STOCK.

Bedding Cattle.—Instead of wasting straw in bedding cattle, it would be much better to pass it through their bodies. If straw must be used for litter, let it be employed as economically as possible. Good substitutes, wholly or in part, for straw bedding may be found in sawdust, ashes, tan and ferns. Leaves of trees if procurable in quantity constitute an excellent litter.

SECTION II.

THE SHEEP.

THE management of sheep varies greatly—depending upon the breeds of the animal, the localities in which they are reared and fattened, and various economic conditions. The tupping season varies of course with the country: in Ireland it commences about the middle of September and lasts for two months; in England and parts of Scotland, the season is about a month earlier. The best kinds of sheep admit of being very early put to breed. Both ram and ewe are ready for this purpose when about fifteen months old. One ram is sufficient for about 80 ewes. The breeding flock should be in a sound, healthy condition, and the ram ought to be as near perfection as possible. The condition of the sire ought to be good, but at the same time it is not desirable to have him over fat. The more striking indications of good health in the sheep are dry eyes, red gums, sound teeth, smooth, oily skin, and regular rumination. The color of the excreta should be natural.

Breeding Ewes.—After the tupping season, which generally lasts for a month, the sheep are usually put on a pasture, which need not be very rich. In cold situations ample shelter should be afforded to the breeding flocks; and in severe weather they should, if possible, be removed to sheds. When snow covers the ground, the animals must be supplied with turnips, or cooked food of some kind. At such time a little oil-cake will be found very useful.

Yeaning.—In March the yeaning season sets in; and as this time approaches, the food of the animals should be improved, and the greatest care must be taken of them. The shepherd should be unceasing in his watchfulness, frequently examining every individual animal. The lambing, if possible, ought to take place in sheds, or some covered place.

Rearing of Lambs.—Delicate lambs require great care. Very weak ones often require to be hand fed. Should a mother die, her offspring may be placed with another ewe; on the other hand, should a lamb perish, its mother may be appointed to rear one of another ewe's twins (if such be available). The ram lambs, not intended for breeding purposes, are subjected to a necessary mutilation when they are about three weeks old. If this operation be performed later, there is great danger that fatal inflammatory action may set in; on the other hand, a lamb much younger than three weeks is hardly strong enough to bear the pain of the operation. The tails of the lambs are shortened about the same time; but it would be better in the case of the rams not to perform both operations on the same day. These operations are best performed during moist or cloudy weather; if they must be done on frosty or stormy days, the lambs should be kept under shelter for two or three days, as otherwise the cold might induce inflammation. The lambs remain with their mothers for about four months, after which they are weaned, and put upon a good pasture. When the herbage is poor, oil-cake, say $\frac{1}{4}$ lb. daily, or some other nutritious food, should be used to supplement it. During the summer and part of the autumn the young stock, as a rule, subsist upon grass; but many flockmasters give them other kinds of food in addition. As winter approaches, the young sheep on tillage farms receive soft turnips, and sometimes a little hay or straw. The allowance of oil-cake may be increased to $\frac{1}{2}$ lb., or if corn be cheap, it may be substituted for the oil-cake. After Christmas Swedish turnips are used.

Mr. Mechi gives the following information on the subject of rearing lambs during a season when roots are scarce:—

Two hundred lambs, which cost 22s. 6d. each on September 12th, were kept on leas and stubble until November 3rd, then on turnips until December 19th, when fifty of them were drafted to another flock getting a little cotton-cake. On the 3rd of February fatting commenced with linseed-cake in addition to cut Swedes. On the 7th of April the fifty tegs were put on rye with mangels, and they were sold on the 4th of May at 61s. each.

The remaining 150 lambs were wintered as stores at little cost, on inferior turnips uncut; they were put on rye from March 8th till May 4th, when they were valued at 48s. each.

The district just referred to became so exhausted of its stock, that at some of the later fairs the number of lambs and of ewes exhibited was less than one-fourth of the average. But in Essex, on six adjoining farms, including that from which I write, the number of sheep wintered has been greater than these heavy lands ever carried before. This has been effected by the extension of a system of management often practised on heavy land, that of eking out a scanty supply of green food by a liberal allowance of straw, chaff, and grain; which happily were good in quality, as well as plentiful and low in price in 1864.

By these means we were enabled last winter to keep 1,500 sheep on about 650 acres of arable, and 350 acres of dry upland pasture—chiefly park surrounding a mansion. The arable land does not very well bear folding in winter, as a preparation for spring corn. Neither climate nor soil are favorable to turnips, and notwithstanding our efforts in assisting Nature, our crops of turnips, rape, or Swedes, are never first-rate, and sometimes very bad. Strong stubbles, good beans, clover-seed, and mangel, are the specialities of the locality, and they indicate heavy land, corn-growing, and yard-feeding. Sheep have been generally " conspicuous by their absence," though even the heavy-land farmer is glad to winter a yard of them instead of cattle, that he may keep some, at least, of the stock that pays best.

In the autumn of 1864 our root crops consisted of some white turnips and rape, eaten by the ewes in September, and of a very bad crop of mangel, the whole of which was reserved for the ewes at lambing-time. In this predicament we wintered about 1,000 half-bred lambs, more than 400 ewes, and some fatting sheep. All, except the fatting sheep, were folded on the stubbles, and allowed a daily run on the park of about an hour for each flock. The freshest grass was reserved for the ewes, and a very meagre bite remained for the lambs; in fact, except for a few weeks in autumn, the parks afforded them little or nothing except exercise and water.

The flocks were divided between three separate farms, and their food was prepared at the respective homesteads. The treatment was in every

THE SHEEP.

respect similar; we shall therefore only notice in detail the management at one farm.

The following details are taken from our "Live Stock Book:"—

EXTRACTS FROM STOCK BOOK.

Lambs.

Payments.	£	s.	d.	Remarks.
November 4th, 1864.				
352 lambs, cost at date, 30s. 9½d. each	542	2	3	Total cost of keeping 352 lambs for 24 weeks, £298 4s. 3d.
Cost of keeping 24 weeks to April 21, 1865:—				Cost per head, 16s. 11d.
Corn and cake, as per granary book	245	16	9	Cost, food only, 14s. 11d.
Cutting 25 tons of chaff, at 6s.	7	13	0	Value of the manure, reckoned at one-fifth the cost of the corn and cake, £49 3s. 4d.
Grinding 96 qrs. 6 bshls. of corn, at 9d.	3	12	6	
Attendance, at 19s. 10d. per week	23	16	0	Cost of the lambs, per head, £2 7s. 8d.
Horse labor, at 6s. per week	7	4	0	
Coal, 3s. 2d. per week	3	16	0	Value of manure, per head, 2s. 10d.
Use of 21 troughs, at 3d. each per month	1	11	6	No charge made for the straw-chaff eaten on the land.
Use of 180 hurdles, at 1d. each per month	4	10	0	
1½ cwt. of rock salt	0	4	6	
	£840	6	6	

The tegs would probably have been sold at a profit in April; they were, however, put on grass and clover, and were fattened in the summer.

September 29th.—352 lambs in the parks, on a little cotton-cake and some oats, until November 4th, when they were folded on a wheat stubble. Gave them 5 bushels of meal daily, mixed with 468 lb. of straw chaff. Cost 3¼d. each per week for meal.

December 20th.—Increased the food to 6¼ bushels of meal and 1 bushel of oil-cake.

December 18th.—

	lb.
2¾ bushels of maize crushed and boiled	143
4½ bushels of mixed meal	200
1 bushel of oil-cake	50
	393

Cost 5½d. per week for corn and cake; chaff, 2¼ lb. each, between these and the ewes, the lambs eating rather less than 2 lb. each.

Eight pounds of rock-salt licked up by the 352 lambs per week.

January 23rd.—The food was increased to 7½ bushels of meal, 2 bushels of oil-cake, and 2 bushels of rape-cake.

Mixture of Corn.		Cost per stone (14 lb.)		
			s.	d.
Wheat	4 parts.	Wheat	1	0
Barley	4 ,,	Barley	0	10
Oats	2 ,,	Oats	1	0
Maize	4 ,,	Maize	0	10
		Oil-cake	1	4¼
		Rape-cake	0	9

Sheep Feeding.—In Ireland sheep are often exclusively fed on grass; but in most cases the addition of other food is desirable, and more especially is it necessary during winter. When confined to roots, sheep, on an average, consume about 26 lbs. daily, unless when under shelter, which diminishes the quantity by from five to ten per cent. Some sheep on which Dr. Voelcker experimented were fed as follows:—

	lbs.	ounces.
Mangel wurzel	19	8
Chopped clover hay	1	3/10
Linseed cake	0	4 8/10
Total	20	15 38/100

On this diet four sheep were maintained from the 22nd of March until the 10th of May, a period of forty-seven days. The weights were as follows:—

	22nd Mar.	10th May.	Gain.
No. 1	153	170½	17½
No. 2	134	151½	17½
No. 3	170	187	17½
No. 4	136	155	19

This experiment shows that the sheep can increase in weight on a daily allowance of food, much less than is usually given to them; but it will be found that growing sheep will usually consume a greater quantity of food than that used by Dr. Voelcker's fattening animals.

Sheep washing is performed before the animal is shorn. It is a process which should never be neglected, as dirty wool is certain to bring a less price than the same quality would if clean. After being washed, sheep should be kept in dry pasture for about ten days in order to allow the loss of yolk removed by the washing to be repaired; they will then be in proper condition for the shearer.

Sheep Dips are used for the purpose of removing parasites from the animal's skin. They often contain arsenic, or bichloride of mercury (corrosive sublimate), which are very objectionable ingredients. The glycerine sheep dip, prepared by Messrs. Hendrick and Guerin, of London, is a safe mixture, as it is free from mineral poisons, whilst the tar substances which it includes, act as a powerful cleanser of the skin, without injuriously affecting the yolk of the wool.

SECTION III.

THE PIG.

IN the breeding of pigs, as in the breeding of other kinds of stock, great care should be taken in the selection of both sire and dam. A good pig should have a small head, short nose, plump cheek, a compact body, short neck, and thin but very hairy skin, and short legs. The black breed is considered to be more hardy than the white; and pure—all black or all white—colors as a rule indicate the purest blood.

The sow should not be bred from until she is a year old, and the boar especially should not be employed at an earlier age. Although one boar is sometimes left with forty pigs and even a greater number, he will not be able to serve more than a dozen about the same time, if vigorous progeny be expected. The sow's regular period of gestation is 113 days; she can have two litters a year, and in each there are from five to fourteen young. Moderate sized litters are the best, the young of very numerous ones being often weakly. The best time to rear young pigs is during the warm or mild parts of the year.

During gestation the sow should be liberally fed, but not with excessive amounts. The food at this time should rather excel in quality than in quantity; but so soon as she begins to nurse, her allowance must be increased, and may be rendered more stimulating. For a week or so before farrowing, the sow ought to be kept alone. Its sty should not be too small—not less than 8 or 10 feet square—for pigs require good air in abundance as well as other animals.

The straw used for litter should neither be too abundant nor too long; in the latter case some of the young might be covered by it, and escaping the notice of the sow, might unconsciously be crushed by the latter. If the young are very feeble, it may become necessary to hand-feed them. Some sows eat their young: and when they have this habit, the better plan is to cease breeding from them; for it appears to be incurable. After parturition some bran and liquid or semi-liquid food should be given to the sow.

Young Pigs subsist exclusively on their mother's milk but for a short time. In two or three weeks they may receive skimmed or butter-milk from the dairy. At a month old such of them as are not designed for breeding purposes may be subjected to the usual mutilations; and at from five to six weeks old the young are weaned, and converted into *stores*.

Store Pigs, when young, are best fed upon skimmed milk, oatmeal, and potatoes, in a cooked state. When they are approaching three months old, they may be supplied with raw food, if the weather be warm; but in winter, cooked and warm food will be found the more economical. Cabbages, roots, potatoes, and all kinds of grain that are cheap are used in pig feeding. The number of meals varies from six or seven in the case of very young animals, to three in the case of those nearly ready for fattening. Store pigs should be allowed a few hours' exercise daily in a paddock, or field, or at least in a large yard.

The dietaries of store pigs vary greatly, for these animals being omnivorous readily eat almost every kind of food. Mr. Baldwin, of Bredon House, near Birmingham, an extensive pig breeder, gave (in 1862) stores the following allowance:—At three months old, a quart of peas, Egyptian beans, or Indian corn. He considered English beans to be too *heating* for young pigs. The animals were allowed the *run* of a grass field. On this diet the stores were kept until they were eight months old (increasing at the average rate of five pounds per week), after which they were allowed an extra half-pint of corn. He calculated the weekly cost as follows:—Dry food, 1s.; grass, 2d.; man's time, 1d.; total, 1s. 3d. These results yielded a profit of 1s. per week per pig, pork being at the time 6d. per lb. Some feeders give young store pigs half-a-pint of peas, mixed with pulped mangel, and the quantum of peas is gradually increased to one pint per diem. All kinds of food-refuse from the house are welcomed by the pig. Skins, dripping, damaged potatoes, cabbage, &c., may be given to them; but they should not be altogether substituted for the ordinary food-stuffs. Coal-dust, cinders, mortar rubbish, and similar substances are often swallowed by pigs, and sometimes even given to them by the feeder. In certain cases Lawes and Gilbert found that superphosphate of lime was a useful addition to the food of pigs. A little salt should invariably be given, more especially if mangels (which are rich in salt) do not enter into the animals' dietary.

Fattening Pigs.—For some time before store pigs are put up to be fattened, the quality and quantity of their food should be increased, for it is not economy to put a rather lean animal suddenly upon a very fattening diet. The sty should be well supplied with clean litter, and should be darkened. Three feeds per diem will be a sufficient number, and the remains (if any) of one should be removed from the trough before the fresh feed is put into it. The feeding trough (which should be made of iron) should be so constructed that the animals cannot place their fore feet in it. The pig is naturally a clean animal, and therefore it should be washed occasionally, as there is every reason to believe that such a procedure will tend to promote the animal's health. It should be supplied with clean water.

In Stephen's "Book of the Farm," it is stated that two pecks of steamed potatoes, and 9 lbs. of barley-meal, given every day to a pig weighing from 24 to 28 stones, will fatten it perfectly in nine weeks. Barley-meal is largely used in England as food for pigs. It is given generally in the form of a thin paste, and in large quantities. Lawes and Gilbert found that 1 cwt. of barley-meal given to pigs increased their weight by $22\frac{1}{2}$ lbs. Indian meal is fully equal, if it is not superior to barley-meal, as food for pigs; and for this purpose it is far more extensively employed in Ireland. Every kind of grain given to pigs should be ground and cooked. In Scotland pigs are often fattened solely on from 28 to 35 lbs. of barley-meal weekly, and mangels or turnips *ad libitum*. Pollard is a good food for pigs, being rich in muscle-forming materials; it is a good addition to very fatty or starchy food. A mixture of pollard and palm-nut meal is an excellent fattening food. Potatoes are now so dear, that they are seldom—unless the very worst and diseased kinds—used in pig feeding. They should never be given raw. The more inferior feeding-stuffs should be used up first in the fattening of pigs, and the more valuable and concentrated kinds during the latter part of the process.

SECTION IV.

THE HORSE.

THE horse is subject to many diseases, not a few of which arise from the defective state of his stable. The best kinds of stables are large and lofty, well ventilated and drained, smoothly paved, and well provided with means for admitting the direct sunlight. The walls should be whitewashed occasionally, and for disinfecting and general sanitary purposes, four ounces of chloride of lime (bleaching powder) mixed with each bucket of whitewash, will be found extremely useful.

Farm horses are kept in stalls, which should not be less than six feet wide, and (exclusive of rack and rere passage) 10 feet long. For hunters and thorough-breds, *loose boxes* are now generally used.

The mare commences to breed at four years, and the period of gestation is 340 days. She may be worked until within a fortnight of the time at which parturition is expected to occur. After foaling, the mare should be turned into a grass field (unless the weather is severe) and kept there idly for three or four weeks.

Foals are kept with their mothers until they are about five or six months old: after weaning, their food must be tender and nutritious—well bruised oats, cut hay, bean or oatmeal mashes; carrots are very suitable.

Working horses are fed chiefly upon oats and hay, which undoubtedly are the best foods for these animals, both being rich in muscle-forming materials. Bruised oats are far more economical than the whole grains: and if the animals eat too rapidly, that habit is easily overcome by mixing chopped straw or hay with the grain.

According to Playfair, a horse not working can subsist and remain in fair condition on a daily allowance of 12 lbs. of hay and 5 lbs. of oats. According to the same authority, a working horse should receive 14 lbs. of hay, 12 lbs. of oats, and 2 lbs. of beans.

Beans are a very concentrated food, rich in flesh-formers, and are, therefore, well adapted for sustaining hard-working horses. They are rather *binding;* but this property is easily neutralised by combining the beans with some laxative food. Turnips, carrots, furze, and various other foods are given to the horse, often in large quantities. The following are some among the many dietaries on which this animal is kept:—

Professor Low's formula is, 30 to 35 lbs. of a mixture of equal parts of chopped straw, chopped hay, bruised grain, and steamed potatoes.

The daily rations of horses of the London Omnibus Company, are 16 lbs. of bruised oats, 7½ lbs. of cut hay, and 2½ lbs. of chopped straw.

Stage coach-horses in the United States receive daily about 19 lbs. of Indian meal and 13 lbs. of cut hay.

Mr. Robertson, of Clandeboye, near Belfast, gives the following information on the subject of horse-keeping :—

The year we divide into three periods—October, November to May inclusive, June to September inclusive. During the first period, the horses get about 18 lb. of chaff and 12 lb. of crushed oats and beans ; " 10½ oats and 1½ beans" per head per day. During the second period they get about 15 lb. of hay chaff, 12 lb. of crushed oats and beans, and about 3 gallons of boiled turnips per head per day. During the third period they were turned out to graze during the night. In the day time, whilst in the stable, each animal is allowed about 50 lb. of cut clover, and about 12 lb. of crushed oats and beans per day. The feeding is all under the charge of one person. He uses his own discretion in feeding the animals, though he is not allowed to exceed the quantities named. The horses to which I allude are the same on which the experiments commenced two years ago—six cart horses, one cart pony, and one riding horse. From Sept. 1, 1865, to and including August 31, 1866, the cost of maintaining these horses in good working condition ; keeping the carts, harness, &c., in repair ; shoeing, &c., was as follows :—

	£	s.	d.
Oats, 14 tons, at 16s. per cwt.	112	0	0
Beans, 2 tons, at 18s. per cwt.	18	0	0
Hay, 13 tons, at 30s. per ton	19	10	0
Green Clover	15	0	0
Turnips	5	0	0
Night grazing	18	0	0
Engine, cutting chaff, crushing oats, &c.	7	4	0
Attendance	26	0	0
Blacksmith	12	0	0
Saddler	12	0	0
Carpenter	10	0	0
Five per cent. interest on value, £110	5	10	0
Depreciation in value 10 per cent.	11	0	0
	£271	4	0
Deduct cost of riding horse	35	0	0
	£236	4	0

£33 11s. 10d. per head; if we suppose the available working days to be 300, allowing 13 for wet days, holidays, &c., the daily cost will be 2s. 2¼d.; to this if we add 1s. 8d., the wages of the driver, we shall have a total of 3s. 10¼d. as the cost of a horse, cart, and driver per day. I would only add, in conclusion, that the horses are kept in good working condition; and, as a proof of their good health under this system, I may state that during the past two years we have not had occasion to require the services of a veterinary surgeon.

Musty hay or straw should not be given to horses. Furze is said to be a heating food; but it is very nutritious, and when young, may be given as *part* of the food of the horse.

Boiled turnips and mangels are often given in winter; but they are not sufficiently nutritious to constitute a substantial portion of the animal's diet. Oil-cake is occasionally given to horses; but seldom in larger quantities than 1½ lbs. per diem. On the whole, experience is in favor of occasionally giving cooked food to horses; and the practice meets with the full approval of the veterinarian. To most kinds of food for horses, the addition of one or two ounces of salt is necessary.

In the *Agricultural Gazette* for November 25, 1865, the following instructive tables are given:—

STABLE FEEDING DURING AUTUMN.

No.	Name and Address of Authorities.	Hay.	Oats.	Beans.	Clover, &c.	Weekly Cost.
		lb.	lb.	lb.		s. d.
1	W. Gater, Botley	168	63*	32*	...	12 0
2	W. C. Spooner	112	84	24	...	11 0
3	T. Aitken, Spalding	...	37½	...	ad lib.	7 6?
4	,, ,,	...	37½	35	ad lib.	10 0?
5	T. P. Dods, Hexham.	...	105	...	ad lib.	10 6?
6	,, ,,	ad lib.	105	...	...	10 6?
7	A. Ruston, I. of Ely.	ad lib. ½	84	10	Straw ad lib. ½ Bran. ½ bush.	9 0
8	A. Simpson, Beauly	168	70	14	24 lb. Straw.	10 0
9	H. J. Wilson, Mansfield	...	52½	...	ad lib.	7 3?
10	,, ,,	42	87½	...	ad lib.	9 0

In this table the asterisk (*) means that the grain is crushed or ground.

STABLE FEEDING DURING WINTER.

No.	Name and Address.	Hay.	Oats.	Beans.	Roots.	Sundries.	Straw.	Weekly Cost.
		lb.	lb.	lb.		lb.	lb.	s. d.
1	Professor Low—Elements of Agriculture	56*	56*	...	Potatoes 56†	...	56*	6 6
2	H. Stephens—Book of the Farm ...	112	35	..	Turnips 112	...		6 0
3	J. Gibson, Woolmet —H. Soc. 1850 ...		84	...	217†	Potatoes 217† Barley 42†	112	9 0
4	— Binnie, Seaton ...		70*	28*	243†		ad lib.	11 6
5	— Thomson, Hangingside	...	84	14	336	14	ad lib.	9 6
6	W. C. Spooner, Ag. Soc. Journ., vol. ix.		63	...	42		196	4 9
7	T. Aitken, Spalding, Lincolnshire ...	ad lib. (⅔)	37	35			ad lib. (⅓)	9 0
8	G. W. Baker, Woburn, Bedfordshire.		60*	20*			...	9 8
9	R. Baker, Writtle, Essex	70	42	...			140	5 0
10	J. Coleman, Cirencester	...	84	16	...	...	ad lib.	7 3
11	T. P. Dods, Hexham	...	95	...	56	Linseed	ad lib.	8 0
12	J. Cobban, Whitfield	84*	60*	...	...	3½	ad lib.*	7 3
13	S. Druce, jun., Ensham	112	52	...	Swedes 70		2 bu.*	7 0
14	C. Howard, Biddenham	ad lib. (⅔)	52	17	84 M. Wurzel		ad lib. ⅓*	8 6?
15	J. J. Mechi, Tiptree.	49*	70*	...	210	...	ad lib.*	7 6
16	W. J. Pope, Bridport	2*	84	...	...	...	ad lib.	9 0?
17	S. Rich, Didmarton, Gloucestershire ...	168	63	...		Grains 2 bush.	ad lib.	10 8
18	H. E. Sadler, Lavant, Sussex	140	84	...	...	...	...	9 9
19	J. Morton, Whitfield Farm	...	126	...	Carrots 350	...	ad lib.	10 9
20	E. H. Sandford, Dover	56	42	...	...	Bran 12	ad lib.	5 6
21	A. Simpson, Beauly, N.B.	...	49	7	105	Tail Corn 21	ad lib.*	5 6
22	H. J. Wilson, Mansfield	42	52½	...	...	Bran 21	ad lib.	6 6?
23	F. Sowerby, Aylesby, North Lincolnshire	112	28	Cut Oat Sheaf.		...	ad lib.*	8 0?

Where an asterisk (*) is attached to any item, it is to be understood that the corn has been bruised or ground, or the hay or straw has been cut into chaff. Where a dagger (†) is appended, the article so marked has been boiled or steamed. A mark of interrogation (?) indicates that the result so marked is uncertain, owing to some indefiniteness in the account given.

On feeding horses with pulped roots, Mr. Slater, of Weston Colville, Cambridgeshire, says :—

I give all my cart horses a bushel per day of pulped mangel, mixed with straw and corn-chaff. I begin in September, and continue using them all winter and until late in the summer, nearly, if not quite, all the year round, beginning, however, with smaller quantities, about a peck, and then half a bushel, the first week or two, as too many of the young-growing mangel would not suit the stock. I believe pulped mangels, with chaff, are the best, cheapest, and most healthy food horses can eat. I always find my horses miss them when I have none, late in the summer. I give them fresh ground every day. Young store beasts, colts, &c., do well with them.

PART IV.

MEAT, MILK, AND BUTTER.

SECTION I.

MEAT.

No one ought to feel a greater interest in the subject of meat in all its branches than the stock feeder. Just in proportion as this kind of food is agreeable to the taste, easily digestible, and rich in nutriment, will the demand for it increase. The quality of meat is, in fact, a primary consideration with the producer of that article; and he whose beef and mutton are the most tender and the best flavored will make the most profit.

Quality of Meat.—The flesh of herbivorous animals is composed of muscular and adipose (fatty) tissues. The muscles consist of bundles of elastic fibres *(fibrine)*, enclosed in an albuminous tissue formed of little vessels, termed cells, and intimately commingled with water, and a mixture of albuminous, fatty, and saline matters. The leanest flesh (muscles) contains fat, but the latter accumulates in certain parts of the body—often to such an extent as to seriously interfere with the functions of life. The red color of flesh is due to a rather large proportion of blood, which it contains in minute vessels; and the slight acidity of its juice is owing to the presence of *inosinic* acid, and probably of several other acids. The agreeable odour of meat, when it is subjected to the process of cooking, is developed from a complex substance termed *osmazome.*[*]

[*] From two Greek words, signifying odour and soup.

This constituent varies in nature and quantity in the different animals—hence the variety in flavor and odour of their flesh—and its amount increases with the age of the animal. The albumen of the muscles, and their fatty and saline constituents, are digestible; but it is generally believed that the elastic fibres, and the horny cellular tissue which binds them into bundles, are not assimilable. It is more certain that the crystalline substances found in flesh, such as, for example, *kreatine*, are incapable of ministering to the nutrition of animals.

The composition of flesh varies very much—that of a very obese pig containing more than half its weight of fat, whilst in some specimens of "jerked beef," imported from Monte Video, scarcely 5 per cent. of that substance was found. The flesh of a fat ox has on an average the following composition:—

	Per cent.
Water	45
Fatty substances	35
Lean flesh, or muscle	15
Mineral matters	5
Total	100

I have examined for Dr. Morgan several specimens of the corned beef recently prepared in South America, by "Morgan's process." The following were the average results of three analyses:—

	Per cent.
Water	40
Fatty matters	21
Lean, or muscular flesh	27
Mineral matters (chiefly common salt)	12
Total	100

It may not here be out of place to direct attention to the composition of a kind of animal food extensively purchased by the poorer classes, and known under the term of slink veal. It is the flesh of calves that are killed on the first day of their existence, and also, I have reason to believe, that of very immature animals—of calves that have never breathed. The flesh is of a very loose texture naturally, and is still further puffed out by air, which is usually supplied from the lungs of the operator. This kind of meat, though regarded as a delicacy by some people, is not held in much estimation, otherwise its price would be higher than it is. It is at present sold at about 4d. or 5d. per pound, sometimes even at a lower rate. Apart from the disgusting process of "blowing" veal, so generally adopted, the use of this food is extremely objectionable, owing to its great tendency to produce diarrhœa. To the truth of this assertion every physician who has studied the subject of dietetics can testify. I have analysed a specimen of it (purchased from a person who admitted that it was part of a calf a day old), and obtained the following results:—

100 parts contain—

	Per cent.
Water	72·25
Fat	6·17
Lean flesh	18·46
Mineral matter	3·12
Total	100·00

I believe that a large portion of the lean flesh is indigestible; and altogether I may safely say of this kind of meat that it is, especially during the prevalence of cholera, an unsafe article of diet. Of course these observations do not apply to *fed* veal, the only kind which respectable butchers, as a rule, offer for sale.

Young meat is richer in soluble albumen and poorer in fibrine and fat than the matured flesh of the same animal. The flesh of the goat contains *hircic* acid, which renders it almost uneatable, but this substance is either altogether absent from, or present but in minute proportion in, the well-flavored meat of the kid. The flesh of game contains abundance of osmazome, a substance which is somewhat deficient in that of the domestic fowl.

Owing to the marked individuality which man exhibits in the selection of his food, and to the intimate relationship subsisting between food and the organism it nourishes, it is impossible to arrange the alimental substances in the strict order of their nutritive values. You can bring a horse to the water, but you cannot compel him to drink it; you can swallow any kind of food you please, but you cannot force your stomach to digest it. It is, therefore, vain to tell a man that a certain kind of food is shown by chemical analysis to be nutritious, when his stomach tells him unmistakeably that it is poisonous, and refuses to digest it. In the matter of dietetics Nature is a safer guide than the chemist. Many substances, when viewed only in the light shed upon them by chemical analysis, appear to be rich in the elements of nutrition, yet when they are introduced into the stomachs of certain individuals, they disarrange the digestive organs, and sometimes cause the whole system to go out of order. Every day we see exemplified the truth of the proverb, that "one man's meat is another man's poison." There are persons who relish and readily digest fat pork, and yet they cannot eat a single egg with impunity; others enjoy and easily assimilate eggs, but their stomachs cannot tolerate a particle of fat bacon.

It is not merely the composition of an aliment and its adaptability to the organism which determine its nutritive value—its digestibility and flavor are points which affect it. There are few people in these countries who are disposed to quarrel with beef; but no one would prefer the leg of an elderly milch cow to the sirloin of a well-fed three-year-old bullock: yet if our selection were to be determined by the analysis of the two kinds of beef, we would be just as likely to prefer the one as the other. No doubt the relative tenderness of meats may be ascertained by experiments conducted *outside* the body; but tenderness is not in every case synonymous with easy digestibility. Veal contains more soluble albumen, and is, consequently, far more tender than beef; yet, as every one knows, it is less digestible. It is curious that maturity renders the flesh of some animals more digestible, and that of others less digestible. Flavor has something to do with these differences.

Beef is richer than veal in the agreeably flavorous osmazome, and the flesh of the kid is destitute of the disagreeable odour of the fully-developed goat. The superiority of wild-fowl over the domesticated birds is solely owing to the finer flavor of their flesh.

The habits of animals, and the nature of their food, affect the quality of their flesh. Exercise increases the amount of osmazome, and consequently renders the meat more savory. The mutton of Wicklow, Wales, and other mountainous regions is remarkably sweet, because the animals that furnish it are almost as nimble as goats, and skip from crag to crag in quest of their food. The fatty mutton, with pale muscle, which is so abundant in our markets, is furnished by very young animals forced prematurely into full development. Those animals have abundance of food placed within easy reach; their muscular activity is next to *nil*, and the result is, that their flesh contains less than its natural proportion of savory ingredients. It is the same with all other animals. The flesh of the tame rabbit is very insipid, whilst that of the wild variety is well flavored. Wild fowls cooped up, and rapidly fattened, lose their characteristic flavor; and when the domesticated birds become wild their flesh becomes less fatty, and acquires all the peculiarities of game. Ducks, whether wild or tame, ordinarily yield goodly meat; but the flesh of some of those that feed on fish smacks strongly of cod-liver oil. Birds which subsist partly on aromatic berries assimilate the odour as well as the nutriment of their food. The flesh of grouse has very commonly a slight flavor of heather. Foster states that in Tahiti pigs are fed upon fruit, which renders their fat very bland and their flesh like veal. Animals subjected to certain kinds of mutilation fatten more rapidly than they do in their natural state. Capons increase in weight more rapidly than cocks, poulards than hens, bullocks than bulls, and cows deprived of their ovaries than perfect cows. Why it is that the flesh of mutilated animals should be fatter and more tender than that of whole animals, we know not; we only know that such is the fact. The hunting of animals renders their flesh more tender; the cause assigned is, that the great exertion of the muscles liquefies their fibrine, which is the toughest of their constituents. The meat of animals brought very early to maturity is seldom so valuable as the naturally developed article. Lawes and Gilbert state that portions of a sheep that had been fattened upon *steeped* barley and mangels, and which gave a very rapid increase, yielded several per cent. less of cooked meat, and lost more, both in dripping and by the evaporation of water, than the corresponding portions of a sheep which had been fed upon *dry* barley and mangels, and which gave only about half the amount of gross increase within the same period of time.

Although the digestibility and flavor of meat (and of every other kind of food) affect its nutritive value, these points are in general of far less importance than its composition. Potatoes are not so nutritious as peas, because they contain a smaller amount of fat and flesh-formers; but they are more digestible. Fish contains less solid matter than flesh, and is less nutritious, yet a cut of turbot will be, in general, more easily digested than an equal weight of old beef. The fact is, that digestibility and flavor are only of great importance to dyspeptic persons. In the healthy digestive organs a pound weight of (dry) food of inferior flavor and slow digestibility will be just as useful as the same weight of well-flavored and easily assimilable aliment, provided all other conditions be alike. If the food be eaten with a relish, and tolerated by the stomach, its digestibility will not, except in extreme cases, affect in a very sensible degree its nutritiveness.

Were one question in animal nutrition satisfactorily answered, it would then be comparatively easy to arrange aliments in the order of their nutritive value. That question is—What are the proper relative proportions of the fat-forming and flesh-forming constituents of our food? It is constantly urged, that the food of the Irish peasantry contains an excess of the fat-forming materials in relation to the muscle-forming substances; and the remedy suggested is, that their staple article of food—potatoes—should be supplemented with flesh, peas, and such like substances, in which, it is supposed, the elements of nutrition are more fairly balanced. In potatoes, the proportion of fat-formers (calculated as fat) is about five times as much as that of the flesh-formers; but these principles exist in the same relative proportions in the fat bacon with which the potato-eater loves to supplement his bulky food. In bread we find the proportion of fat-formers to be only $2\frac{1}{2}$ times as much as that of the flesh-formers, whilst, according to Lawes and Gilbert, the edible portion of the carcass of a fat sheep contains $6\frac{1}{2}$ times as much fat as nitrogenous (flesh-forming) compounds. It is evident, then, that meat such as, for example, the beef recently imported from Monte Video, from which the fatty elements of nutrition are almost completely absent, cannot be a suitable adjunct to a farinaceous food.

There is evidence to prove that in the animal food consumed by the population of these countries, the proportion of fatty to nitrogenous matters is greater than in the seeds of cereal and leguminous plants, and but little less than in potatoes. "It would appear to be unquestionable," say Lawes and Gilbert, "therefore, that the influence of our staple *animal foods*, to supplement our otherwise mainly farinaceous diet, is, on the large scale, to *reduce*, and *not to increase*, the relation of the *assumed* flesh-forming material to the more peculiarly respiratory and fat-forming capacity, so to speak, of the food consumed." It must be remembered, too, that the fat *formers* are ready *formed* in animal food, whereas they exist chiefly in the form of starch, gum, sugar, and such-like substances in vegetables. According to theory, $2\frac{1}{2}$ parts of starch are equivalent to, *i.e.*, convertible into, 1 part of fat; but it is not certain whether the force which effects this change is derivable from the $2\frac{1}{2}$ parts of starch, or from the destruction of tissue, or of another portion of food. If there be a tax on the system in order to convert starch into fat, it is evident that $2\frac{1}{2}$ parts of starch, though convertible into, are not equivalent in nutritive value to one part of fat.

It is quite certain that millions of healthy, vigorous men have subsisted for years exclusively on potatoes; but it is no less clear that a diet of meat and potatoes enables the laborer to work harder and longer than if his food were composed solely of potatoes. But we have seen that the relation between the flesh-forming and fat-forming elements is nearly the same in both potatoes and meat; so that the superiority of a meat or mixed diet cannot be chiefly owing, contrary to the generally received opinion, to a greater abundance of flesh-forming materials. As the proportion of flesh-formers to fat-formers is so much greater in wheaten or oaten bread than in potatoes, and as peas and other vegetables rich in nitrogenous compounds are practically found to be an excellent supplement to potatoes, it is probable that the latter may be somewhat relatively deficient in flesh-forming capacity. It is, however, in all probability the great bulk of a potato diet, and its total want of ready formed fat, that render the addition to it of animal food so very desirable. The concentrated state in which the ingredients of flesh exist, the intimate way in which they are intermixed, their agreeable flavor, and their (in general) ready and almost complete digestibility, appear to be the principal points in which a meat diet excels a vegetable regimen. There may be others, which, though less evident, are, perhaps, of equal importance. At all events, the general experience of mankind testifies to the superiority of a mixed animal and vegetable diet over a purely vegetable one.

Is very Fat Meat wholesome?—The enormous and rapidly increasing demand for meat which characterises the food markets of these days, has reacted in a remarkable manner upon the nature of the animals that supply it. Formerly the animals that furnished pork, mutton, and beef, were allowed to attain the age of three years old and upwards before they were considered to be "ripe" for the butcher; but now sheep and pigs are perfectly *matured* at the early age of one year, and two-year-old oxen furnish a large quota of the "roast beef of old England." The so-called improvement of stock is simply the forcing of them into an unnatural degree of fatness at an early age ; and this end is attained by dexterous selection and crossing of breeds, by avoidance of cold, by diminishing as much as possible their muscular activity, and lastly, and chiefly, by over-feeding them with concentrated aliments.

Every one knows that a man so obese as to be unable to walk cannot be in a healthy state; yet many feeders of stock look upon the monstrously fat bulls and cows of cattle show prize celebrity as normal types of the bovine tribe. It requires but little argument to refute so fallacious a notion. No doubt it is desirable to encourage the breeding of those varieties of animals which exhibit the greatest disposition to fatten, and to arrive early at maturity; but the forcing of individual animals into an unnatural state of obesity, except for purely experimental purposes, is a practice which cannot be too strongly deprecated. If breeders contented themselves with handing over to the butcher their huge living blocks of fat, the matter would not perhaps be very serious ; but, unfortunately, it is too often the practice to turn them to account as sires and dams. Were I a judge at a cattle show, I certainly should disqualify every extremely fat animal entered for competition amongst the breeding stock. Unless parents are healthy and vigorous, their progeny are almost certain to be unhealthy and weakly ; and it is inconceivable that an extremely obese bull and an unnaturally fat cow could be the progenitors of healthy offspring. We should by all means improve our live stock ; but we should be careful not to overdo the thing. If we must have gaily-decked ponderous bulls and cows at our fat cattle exhibitions, let us condemn to speedy immolation those unhappy victims to a most absurd fashion; but in the name of common sense let us leave the perpetuation of the species to individuals in a normal state, whose muscles are not replaced by fat, whose hearts are not hypertrophied, and whose lungs are capable of effectively performing the function of respiration.

Mr. Gant, in a small volume* devoted wholly to the subject, describes the serious functional and structural disarrangements which over-feeding produces in stock. He found the heart of a one-year old Southdown wether, fattened according to the *high-pressure system*, to be little more than a mass of fat. In several other young, but so-called "matured" sheep, he found more or less fatty degeneration of the heart, and extensively spread disease of the liver and of the lungs. A four-year old Devon heifer, exhibited by the late Prince Consort at a Smithfield show, was found to be in a highly diseased state. It was slaughtered, and of course its flesh sold at a high price as "prize beef," but its internal organs came into Mr. Gant's possession. The substance of both ventricles of the heart had undergone all but complete conversion into fat; one of its muscles was broken up, and many of the fibres of the others were ruptured. In another animal the muscular fibres of the heart had given way to so great an extent that if the thin lining membrane (*endocardium*) had burst, death would have instantly ensued. The slightest exertion was likely to cause this catastrophe; but, fortunately enough in this case, the animal was not capable of exertion, for though under three years of age, it weighed upwards of 200 stones: this animal had received for some time before its exhibition, the liberal allowance of 21 lbs. of oil-cake (besides other food) per diem. "A pen of three pigs," says Mr. Gant, "belonging to his Royal Highness the Prince Consort, happened to be placed in a favorable light for observation, and I particularly noticed their condition. They lay helpless on their sides, with their noses propped up against each other's backs, as if endeavouring to breathe more easily, but their respiration was loud, suffocating, and at long intervals. Then you heard a short catching snore, which shook the whole body of the animal, and passed with the motion of a wave over its fat surface,

* "A New Inquiry, fully illustrated by coloured engravings of the heart, lungs, &c., of the Diseased Prize Cattle lately exhibited at the Smithfield Cattle Club, 1857." By Frederick James Gant, M.R.C.S. London, 1858.

which, moreover, felt cold. I thought how much the heart under such circumstances must be laboring to propel the blood through the lungs and throughout the body. The gold medal pigs of Mr. Moreland were in a similar condition, if anything, worse; for they snored and gasped for breath, their mouths being opened, as well as their nostrils dilated, at each inspiration. From a pig we only expect a grunt, but not a snore. These animals, only twelve months and ten days old, were marked '*improved* Chilton breed.' They, with their fellows just mentioned, of eleven months and twenty-three days, had early come to grief. Three pigs of the black breed were in a similar state, at seven months three weeks and five days, yet such animals 'the judges highly commended.'"

Dr. Brinton denies the accuracy of several of Mr. Gant's statements relative to the structural changes in the muscles of obese animals; but I do not think that he has succeeded in disproving the principal assertions made by the latter.

There is conclusive evidence to prove that one of the effects of the present mode of fattening beasts is disease of the internal organs of the animals; but it is by no means certain that the flesh of those diseased animals is as unwholesome food as some writers assert it to be. The flesh of an over-fattened animal differs from that of a lean, or moderately fat one, in containing an exceedingly high proportion of fat; but it has not been proved that the fat of prize animals differs from the fat of lean kine, or that it is less wholesome or nutritious. Be the flesh of those exceedingly fat animals unwholesome or not, there are thousands, ay, millions of persons, to whom its greasy quality renders it peculiarly acceptable; and as for those who dislike fat—they do not usually invest their money in the flesh of prize sheep or oxen. At the same time, it must not be understood that all, or even a large proportion of fully matured stock is in a diseased state; though in most of them the vital and muscular powers are undoubtedly exceedingly low.

There is no doubt but that sheep and oxen, from three to

five years old, moderately fat, and fairly exercising their locomotive powers, furnish the most savory, and, perhaps, the most nutritious meat : but if such were the only kind of meat in demand, it may be fairly doubted that the supply would be equal to it. The produce of meat in these countries has been rapidly increasing for many years past; and the weight of meat annually supplied from a given area of land is now from 80 to 100 per cent. greater than it furnished thirty or forty years ago. It is chiefly by means of the so-called forcing system that the produce of meat has been so considerably increased. If this system were abandoned, the production would be greatly diminished, and the consequently high price of the article would place it beyond the reach of the masses of the population. Besides, it has not been proved that the flesh of the animals brought early to maturity is much inferior, except somewhat in flavor, to the meat of three-year-old beasts. There is, no doubt, plenty of unwholesome meat offered for sale, but it is that of animals which were affected by diseases as likely to attack the young as the old. On the whole, then, we may say of the improved system of fattening stock, that it produces a maximum amount of meat on a given area of land ; that the meat so produced is, except in rare cases, perfectly wholesome ; that it is capable of supplying the ingredient—fat—which is almost wholly absent from a vegetable diet; and, finally, that it places animal food within the reach of the working classes.

Diseased Meat.—The losses occasioned to stockowners by the diseases of live stock are far greater than is generally supposed. It has been calculated that in the six years ending 1860, the value of the horned stock lost by disease amounted to £25,934,650. Pleuro-pneumonia was the chief cause of these losses. Exclusive of the enormous losses occasioned by the ravages of the rinderpest, the annual loss by disease in live stock in these countries for some years past cannot be much under £6,000,000 sterling.

Whether it is owing to the somewhat abnormal condition under which the domesticated animals are placed, or to causes which operate upon them when in a state of nature, it is certain that they are remarkably prone to disease. It is extremely difficult to get a horse six years old that is not a roarer or a whistler, or "weak on his pins," or in some way or other unsound. Oxen, sheep, and pigs have almost as many maladies afflicting them as human flesh is heir to, notwithstanding the short period of life which they are permitted to enjoy.

It is a very serious question whether or not the flesh of animals that have been killed while they are in a diseased condition is injurious to health. The opinions on this point are conflicting, but the majority of medical men believe that the flesh of diseased animals is not wholesome. There are certain maladies which obviously render meat unsaleable, by causing a sensible alteration in its quality. For example, blackleg in cattle and measles in the porcine tribe render the flesh of these animals, as a general rule, unmarketable, or nearly so. But there are very serious diseases—often proving rapidly fatal—which, whilst seriously affecting certain internal organs, do not palpably deteriorate the quality of the flesh. In such cases are we to rely upon the evidence of our mere senses in judging of the wholesomeness of the meat? If we find beef possessing a good color and odour, and firm to the touch, and *appearing* to be in every respect healthy flesh, are we under such circumstances to take it for granted that it must be healthy? This is a very important question, involving as it does the interests of both the producers and consumers of animal food. If the flesh of all diseased animals be unwholesome, a very large number of oxen now sold whilst laboring under pleuropneumonia should not be sent into the market. This, of course, would be a heavy loss to the stockowner, but a still heavier one to the meat consumer; because, if there were fewer animals for sale, the price of meat would ascend, in obedience to the law of supply and demand. The whole question is, then, well worthy of being considered in the most careful, unbiassed, and scientific manner; for at present it is in a state which is the reverse of being satisfactory.

A large proportion of the animals conducted to the shambles is in a diseased condition. Professor Gamgee estimates it at no less than one-fifth. Dr. Letheby, food analyst to the Corporation of London, condemns weekly about 2,000 pounds weight of flesh; but as his jurisdiction is limited to the "City," which contains a population of only about 114,000, the 2,000 pounds of diseased meat are probably only about 1-30th of the quantity exposed for sale within the whole area of the metropolis. Making an estimate of the most moderate kind, we may assume that 30,000 pounds weight of bad meat are weekly offered for sale in London—*three million pounds weight annually.*

Many persons have been affected with dysentery and choleraic symptoms after partaking of butcher's meat of apparently the most healthy kind. The meat has often been subjected to minute chemical and microscopical examination, but no poison has been discovered. But these cases are becoming so frequent that they are exciting uneasiness, and demand an exhaustive investigation. The unskilful persons who officiate in the capacity of "clerks of the market" and inspectors of meat can only judge of the quality of flesh that is obviously inferior to the eye, nose, or touch; but are there not cases where the flesh may appear to be good, and yet contain some subtle malign principle? It is an ascertained fact that young or "slink" veal very frequently gives rise to diarrhœa, more especially when that disease is epidemic. Dr. Parkes, in his celebrated work on Hygiene, page 162 (second edition), states that "the flesh of the pig sometimes produced diarrhœa—a fact I have had occasion to notice in a regiment in India, and which has often been noticed by others. The flesh is, probably, affected by the unwholesome garbage on which the pig feeds." Menschell states that 44 persons were afflicted with anthrax after eating the flesh of oxen affected with carbuncular fever. Dr. Kesteren, in the *Medical Times* for March, 1864, mentions a case where twelve persons were affected with choleraic symptoms after the use of pork not obviously diseased. At Newtownards, county of Down, several persons died after eating veal in which no poisonous matter of any kind could be detected. One instance has come under my own notice where a man, two dogs, and a pig died after eating the flesh of an animal killed whilst suffering from splenic apoplexy. Several butchers have lost their lives in consequence of the blood of diseased animals being allowed to come in contact with abrasions or recently received wounds on their arms. The flesh of over-driven animals is stated by Professor Gamgee to produce a most serious skin disease, although the meat appeared to be perfectly healthy. The Belgian Academy of Medicine has decided that the flesh of animals suffering from carbuncular fever is unwholesome, and its sale in that country is prohibited.

Many persons have died in Germany and a few in England from a disease produced by eating pork containing a small internal parasite termed *trichina spiralis*. I have recently met with a case of *trichiniasis* in the human subject. The body of the unfortunate person—who had been an inmate of the South Dublin Union Workhouse—was found to contain thousands of the trichinæ. In Iceland a large proportion of the population suffers from a parasitic disease traceable to the use of the flesh of sheep and cattle in which flukes abound.

Pleuro-pneumonia is in this country the disease which most frequently affects the ox. It is probable that about 5 per cent. of these animals sold in Dublin are more or less affected by this malady. There are two forms of pleuro-pneumonia—the sporadic, or indigenous, and the foreign, or contagious. It is the latter form which has become the scourge of the ox tribe in this country, though unknown here until the year 1841, when it appeared as an epizoötic, and carried off vast numbers of animals.

The contagious pleuro-pneumonia is an extremely severe inflammatory disease, and is produced—not in the same way that common pleuro-pneumonia is, by exposure to excessive cold, &c.—but by a blood poison received from an infected animal. In the congestive stage of the disease there is no structural alteration in the organs of the animal, and if well bled its flesh might (probably) be safely eaten; but when a large portion of the lungs becomes solidified, and rendered incapable of purifying the blood, is it not doubtful, to say the least, that the blood or flesh is perfectly wholesome? The blood, during the life of the animal, is in a state of fermentation; there is extreme fever, and the animal presents all the characteristic symptoms of acute disease. On being killed, the flesh, if the disease be of a fortnight's duration, will usually be extremely dark, but in a less advanced stage of the malady the flesh will generally present a healthy appearance. Is it really so? That is the question which science has to determine. Going upon a broad principle, I can hardly conceive that so serious a disease as pleuro-pneumonia does not injuriously affect the quality of the flesh. It is no argument to say that thousands consume such flesh, and yet enjoy good health. Millions of people drink water and breathe air that are extremely impure, and yet they do not speedily die. It is one thing to be poisonous, another to be unwholesome. The flesh of animals killed whilst suffering from lung distemper is not directly poisonous, but who can prove that it is not, like bad water, unwholesome?

As analyst to the city of Dublin, I am almost daily called upon to inspect meat suspected to be unwholesome; and I have always condemned as being unfit for human food:—

1. Animals slaughtered at the time of bringing forth their young.
2. Oxen affected with pleuro-pneumonia, when pus is present in the lungs, or the flesh obviously affected; animals suffering from murrain, black-quarter, and the different forms of anthrax.
3. Animals in an anæmic, or wasted condition.
4. Meat in a state of putrefaction.

During the present year about 20,000 pounds weight of meat have been seized and condemned in the city of Dublin.

SECTION II.

MILK.

MILK is a peculiar fluid secreted by the females of all animals belonging to the class *Mammalia;* and, being designed for the nourishment of their offspring, contains all the constituents which enter into the composition of the animal body.

The milk of different animals varies very much in color, taste, and nutritive value. That of the cow is a little heavier than water—its specific gravity being, on the average, about 1·030, water being 1·000. It is composed of three constituents—namely, butter, curd, and whey—each of which is also composed of a number of substances. These three constituents are of unequal weight, or specific gravity, and their separation is the chief process carried on in the dairy. The butter is the lightest and the curd is the heaviest constituent.

The following table represents the composition of the milk of different animals:—

COMPOSITION OF THE MILK OF DIFFERENT ANIMALS.

1,000 PARTS CONTAIN—

	Specific Gravity, or Density.	Water.	Solid Ingredients.	Cheesy Matter.	Sugar.	Butter.	Mineral Matter.
Woman	1032·67	889·08	110·92	39·30	43·68	26·66	1·30
Cow	1030	864·20	135·80	48·80	47·70	31·30	6·00
Goat	1033·53	844·90	155·10	35·14	36·91	56·87	6·18
Ewe	1040·98	832·32	167·68	69·78	39·43	51·31	7·16
Mare	1033·74	904·30	95·70	33·35	32·76	24·36	5·23
Ass	1034·57	890·12	109·88	35·65	50·46	18·53	5·24
Bitch	1041·62	772·08	227·92	116·88	15·29	87·95	7·80

Milk examined through a microscope is a colorless fluid, containing a large number of little vesicles, or bags, filled with butter—a mixture of oily and fatty matters. When the milk stands for some time, the globules, being lighter than the other constituents, ascend to the top, and, mixed with a certain proportion of milk, are removed as cream. The curd is termed in scientific parlance *casein*, and is in fresh milk in a state of solution—that is to say, is dissolved in milk in the same way that we dissolve sugar in water. When milk becomes sour, either naturally or by the addition of rennet, it can no longer hold casein in solution, and the curd consequently separates. Casein is the substance which forms the basis of cheese. The substance that remains after the removal of the butter and cheese is called *serum*, or whey, and is composed of a sweetish substance termed *sugar of milk*, and certain saline bodies, termed the ash, dissolved in water.

The butter and the sugar of milk are employed in the animal economy in the production of fat, and are what have been styled by physiologists *heat-producers* and *fat-formers*. The casein resembles the gluten of wheat in composition; it belongs to the class of food substances termed *flesh-formers*. The ash, or mineral part of the milk, is chiefly employed in forming the bones of the young animals it is destined to nourish.

The quality of milk is influenced by the quantity and quality of the food given to the animal. The milk of cows fed on distillery wash, turnip, and mangel tops, coarse herbage, and other kinds of inferior food, is always of inferior quality. Hence it is of great importance that dairy stock be kept in good old pastures in summer, and fed on Swedish turnips, mangelwurzel, and oil-cake during winter. It is true economy to supply dairy cows with abundance of nutritious food; and it should be constantly borne in mind that the milk from two well-fed cows will give more butter than can be obtained from the produce of three badly-fed animals.

The butter is the constituent of milk which is most affected by the nature and amount of the animal's food; and butter is precisely the article which is of the greatest importance to the Irish dairy farmer, as the quantity of cheese prepared in this country is inconsiderable. When, therefore, it is found that a cow pastured on inferior land, or badly fed in the byre, yields a large supply of milk of a high specific quantity (which, however, is rarely the case), it must not be concluded that the result is satisfactory; for if such milk be tested by the lactometer it will certainly be found wanting in butter. The average composition of English milk, according to Way, is:—

Water	87.02
Butter	3.23
Casein	4.48
Sugar of milk	4.67
Ash	0.60
	100.00

In several analyses of milk published by Professor Voelcker, the highest proportion of butter is stated to be 7.62. In that of cows kept on poor and over-stocked pastures less than 2 per cent. was found. I have examined in my capacity of Food Analyst to the City of Dublin several hundred samples of milk, in not one of which have I found the proportion of butter to amount to more than 5.6 per cent. In no sample did I find a higher per-centage of solid matter than 13.15, or (when pure) lower than 12.08. The quality of the food of the milch cow exercises a great influence on the quality and yield of her milk. Aliments rich in fat and sugar favor the production of butter, and augment the supply of milk. Locust-beans, malt, and molasses are good milk-producing foods; but the chief condition in the production of milk rich in butter is simply that the animals which yield it must be fed with abundance of nutritious food. Nor must it be supposed that the richness of milk is due to the smallness of the yield, for whenever the quality of the secretion is inferior, it is almost certain to be deficient in quantity. Those cows which give the richest milk, generally yield the largest quantity.

Yield of Milk.—According to Boussingault, a cow daily yields on the average 10.4 parts of milk per 1,000 parts of her weight. Morton, in his "Cyclopædia of Agriculture," p. 621, states that Mr. Young, a Scotch dairy keeper, obtained 680 gallons per cow per annum. Voelcker found that some common dairy stock gave each of them fifty-two pints of milk per diem, whilst three pedigree cows yielded respectively forty-nine pints.

Professor Wilson gives the following information on this point :—

Our principal dairy breeds are the Ayrshire, the Channel Islands, the Short-horn, the Suffolk, and the Kerry. Some published returns of two dairies of Ayrshire cows give the annual milk produce per cow at 650 and 632 gallons respectively. Three returns of dairies, consisting wholly of Short-horns, show a produce of 540 gallons, 630 gallons, and 765 gallons respectively, or an average of 625 gallons per annum for each cow. In two dairies, where half-bred Short-horns were kept, the yield was 810 and 866 gallons respectively for each cow. In four dairies in Ireland, where pure Kerrys and crosses with Short-horns and Ayrshires were kept, the annual produce per cow was returned at 500 gallons, 600 gallons, 675 gallons, and 740 gallons respectively; or an average, on the four dairies, of 630 gallons per annum for each cow. A dairy of "pure Kerrys" gave an average of 488 gallons per cow, and another of the larger Irish breed gave an average of 583 gallons per head per annum. In the great London dairies, now well-nigh extinguished by the ravages of the cattle disease, these returns are greatly exceeded. The cows kept are large framed Short-horns and Yorkshire crosses, which, by good feeding, bring the returns to nearly 1,000 gallons per annum for each cow kept. The custom in these establishments is to dispose of a cow directly her milk falls below two gallons a-day, and buy another in her place.

The following milk return of one of our best managed dairy farms (Frocester Court) shows the relative produce of cows in the successive years of their milking. The first lot was bought in at two-years old; all the others at three years :—

No. of Cows.	Year of Milk.	Produce per head.
8	1st	317 gals.
15	1st	472 ,,
14	2nd	353 ,,
15	3rd	616 ,,
20	4th	665 ,,
18	5th	635 ,,
9	6th	708 ,,
15	Old	651 ,,

The maximum reliable milk produce that we have recorded was that of a single cow belonging to the keeper of the gaol at Lewes, the details of which were authenticated by the Board of Agriculture. In eight consecutive years she gave 9,720 gallons, or at the rate of more than 1,210 gallons per annum. In one year she milked 328 days, and gave 1,230 gallons, which yielded 540 lbs. of butter, or at the rate of 1 lb. of butter to 22¾ lb. of milk. In the early part of the present year (1866) a return was published of the produce of a cow in a Vermont (U.S.) dairy, which was stated to have given, in the previous year, a butter yield of 504 lbs., at the rate of 1 lb. of butter to 20 lbs. of milk.*

Preserved Milk.—Various plans have been proposed to render milk more portable, and to preserve it sweet for days and even months. Mr. Borden of Connecticut, United States, prepares a concentrated milk by boiling the fluid down in vacuo, at a temperature under 140° Fahrenheit, mixing the resulting solid with sugar, and rapidly placing the compound in tins, which are then hermetically sealed. It is said that solidified milk prepared by this process remains sweet for many months. In France, solidified and concentrated milk are largely prepared; and it is certain that London and other large towns will yet be supplied with milk rendered portable and more stable, by the removal of a large proportion of its water. In many parts of Ireland pure milk could be bought at from 7d. to 8d. per gallon. I do not despair to see factories established in such places for the manufacture of preserved milk as a substitute for the dear and impure fluid sold under the name of milk in London and other large cities. It is stated that solidified milk prepared in Switzerland is now sold in London.

SECTION III.

BUTTER.

History of Butter.—The very general use of butter as an article of food is demonstrated by the familiar saying—" We should not quarrel with our bread and butter"; yet this article,

* Professor John Wilson's Report of the Agricultural Exhibition, Aarhuus, 1867.

now so commonly used throughout the greater part of Europe, was either unknown or but imperfectly known to the ancients. In the English translation of the Holy Scriptures the word butter does certainly frequently occur; but the Hebrew original is *chamea*, which, according to the most eminent Biblical critics, signifies cream, or thick, sour milk. In the 20th chapter of Job the following passage occurs:—"He shall not see the rivers, the floods, the brooks of honey and butter." Now, we can conceive streams of thin cream, but we cannot imagine a river of butter. The oldest mention of butter is found in the works of Herodotus. In the description of the Scythians given by this ancient author, reference is made to their practice of violently shaking the milk of their mares, for the purpose of causing a solid fatty matter to ascend to its surface, which, when removed from the milk, they considered a delicious article of food. Hippocrates, who wrote a little later than Herodotus, describes, but in clearer language, the manufacture of butter by the Scythians; he also alludes to the preparation of cheese by the same people. The word, butter, does not occur in any of Aristotle's writings, and although mention is made of it in the works of Anaxandrides, Plutarch, and Ælian, it is evident that they considered it only in the light of a curious substance, employed partly as an article of food, partly as a medicinal salve, by certain barbarous nations. About the second or third century, butter was but little known to the Greeks and Romans, and there is no reason to believe that it was ever generally used as an article of food by the classic nations of antiquity; it is noteworthy, that the inhabitants of the south of Europe even at the present time use butter in very small quantities, which, indeed, is often sold for medicinal purposes in the apothecaries' shops in Italy, Spain, and Portugal. From the foregoing statements it is evident that the butter manufacture can lay no claim to a classic origin; but that it took its rise in the countries of savage, of semi-civilised, and barbarous nations. It is probable that the Greeks were made acquainted with butter by the Thracians, Phrygians, and

Scythians; and that the knowledge of this substance was conveyed to Rome by visitors from Germany. During the middle ages the practice of butter-making spread throughout Northern, Central, and Western Europe; but in many parts the commodity was very scarce and highly valued, notwithstanding its being almost, if not quite, in a semi-fluid state, instead of possessing the firm consistence of the butter of the present day.

Irish Butter.—Butter is produced in such large quantities in Ireland that, after the home demand has been supplied, there remains a large excess—so considerable, indeed, as to constitute one of the more important of our few commercial staples. The precise quantity of butter which, during late years, has been annually exported from Ireland is unknown. The greater part of the commodity is sent to trans-Channel ports; and, there being no duty on butter in the cross-Channel trade since 1826, we have no means of accurately estimating the amount of our exports to Great Britain. If, however, we refer to the statistics of our commerce for the period beginning in 1787, and ending in 1826, we shall find that the exportation of butter was enormous, and that a large proportion of that commodity consumed by the army and navy was supplied from the dairies of Ireland. During the three years ended on the 5th of January, 1826, the average annual amount of butter exported was as follows :—

	cwts.
To Great Britain ...	... 441,226
To foreign countries ...	... 51,637

Of late years the exportation to foreign and colonial countries has fallen off; still the export trade is very considerable, probably amounting to 450,000 cwts. per annum. During the year 1867, the imports of foreign butter into Great Britain amounted to 1,142,262 cwts.

I have quoted the above statistics for the purpose of demonstrating the great importance of the butter trade to this country. Not only is a large proportion of the agricultural community pecuniarily interested in the production of this article, but the exportation is the chief cause of the commercial prosperity of a city, which, in point of population, ranks third in the kingdom. If butter, then, be an article of so much importance, it is obvious that the greatest care should be taken in its preparation, and that the efforts of both scientific and practical men should be directed towards the best mode of improving its quality. If the principles involved in the production of butter were thoroughly understood, and generally known, I believe that such terms as "seconds," "thirds," and "fourths," would speedily fall into disuse; that there would be only one kind of butter sent into the market; and that the article would always be of the best quality, in other words, "firsts."

Composition of Butter.—The composition and quality of butter depend to a great extent upon the condition of the milk or cream from which it is prepared, and on the skill and cleanliness of the dairy-maid. It consists essentially of fatty and oily matters, but it is always found in combination with casein (cheesy matter) and water. The following analyses, made by Mr. Way, late consulting chemist to the Royal Agricultural Society of England, shows its composition:—

INGREDIENTS PER CENT.

	1.	2.	3.
Fatty matters	82.70	79.67	79.12
Casein	2.45	3.38	3.37
Water	14.85	16.95	17.51

No. 1 analysis shows the composition of a specimen obtained from the well-known Mr. Horsfall's dairy. It was made from raw cream. The other specimens were the produce of a Devonshire dairy, and were prepared from scalded cream. In several specimens of well-made and unsalted Irish butter which I have analysed, I found the proportion of casein or cheesy matter never to exceed 1 per cent., whilst in the analysis above stated the centesimal amount is on the average more than 3 per cent.

The fatty matter is composed of two substances—one, a

solid, termed *margarin;* the other fluid, and styled by chemists *elaine.* The solid fat is identical in composition with the solid fat of the human body. The elaine is peculiar to milk, but it differs very slightly from *olein,* or fluid fat. The relative proportions of the fluid and solid fats vary with the seasons. According to Braconnot, the solid fat forms in summer 40 per cent. of the butter, but in winter the proportion rises to 65. This decrease in the proportion of the liquid fat in winter is the cause of the greater hardness of the butter in that season, which is often incorrectly attributed solely to the cold.

The cheesy and acid matters contained in butter are by no means essential; on the contrary, if it were quite free from them, it might be retained with little or no salt for a very long period without becoming rancid. The cheesy matter contains nitrogen; and nearly all the substances into which this element enters as a constituent are remarkably prone to decomposition. Yeast, and ferments of every kind—gunpowder, fulminating silver, chloride of nitrogen—and almost every explosive compound, contain this element. The cheesy matter is a very nitrogenous body, and in presence of air and moisture not only rapidly decomposes, or decays, itself, but induces by mere contact a like state of decomposition in other substances—such, for instance, as fat, sugar, and starch, which naturally have no tendency to change their state. Bearing the foregoing facts in mind, it is obvious that the chief precautions to be observed in the manufacture of butter are :—Firstly, to separate to as great an extent as practicable the casein from the butter; and, secondly, as in practice a small portion of the curd remains in the butter, to prevent it from undergoing any change—at least for a prolonged period. How these desiderata may best be accomplished I shall now proceed to point out.

The Butter Manufacture.—The theory of the process of churning is very simple. By violently agitating the milk or cream the little vesicles, or bags containing the butter, are broken, and, the fatty matter adhering, *lumps of butter* are formed. The operation of churning also introduces atmospheric air into the milk, which, aided by the high temperature to which the fluid is raised, converts a portion of the *sweet* sugar of milk into the *sour* lactic acid. By the alteration produced in this way in the composition of the milk, it is no longer capable of holding the casein in solution, and the curd therefore separates.

The churn and other vessels in which the milk is placed cannot be kept too clean. No amount of labor bestowed on the scalding and scrubbing of the vessels is excessive. When wood is the material used in the milk-pans the utmost care should be taken in cleaning them, as the porous nature of the material favors the retention of small quantities of the milk. A simple washing will not suffice to clean such vessels. They must be thoroughly scrubbed and afterwards well scalded with *boiling* water. Tin pans are preferable to wooden ones, as they are more easily cleaned, but in their turn they are inferior to glass vessels, which ought to supersede every other kind. Earthenware, lead, and zinc pans are in rather frequent use. The last-mentioned material is easily acted upon by the lactic acid of the sour milk, and is, therefore, objectionable. It is a matter of great importance that the dairy should not be situated near a pig-stye, sewer, or water-closet, the effluvia from which would be likely to taint the milk. It is surprising how small a quantity of putrescent matter is sufficient to taint a whole churn of milk; and as it has been demonstrated that the almost inappreciable emanations from a cesspool are capable of conferring a bad flavor on milk, it is in the highest degree important to remove from the churn and milk-pail every trace of the sour milk. I go further, it is even desirable that no one whose hands have a tendency to perspire should be allowed to manipulate in the dairy; and it should be constantly borne in mind that the dairy-maid's fingers and hot water should be on the most intimate visiting terms.

Butter is made either from cream—sour and sweet—or from whole milk which has stood sufficiently long to become distinctly sour. It is asserted by some makers that butter prepared from whole milk, or from scalded cream, contains a large proportion of curd. If this be true—which I greatly doubt—it is a serious matter, for such butter would speedily become rancid in consequence of the casein acting as a ferment. I believe that experience points to an exactly opposite conclusion. From the results of careful inquiries I feel no hesitation in asserting that the butter should not be made from the cream, but from the *whole milk*. When made from the cream alone it is much more likely to acquire a bad taste, and is generally wanting in keeping qualities. I have no doubt but that in the process of churning the whole milk there is a large amount of lactic acid formed, and a much higher temperature attained, than in the churning of cream; consequently, the separation of caseous matter must be more perfectly effected in the former than in the latter case. It is a mistake to think that there is very little casein in cream: out of 7 or 8 lbs. of thick cream only a couple of pounds of butter are obtainable; the rest is made up of water, casein, and sugar of milk. The yield of butter is greater when the whole milk is churned than when the cream alone is operated upon, and, what is of great importance, the quality of the butter is uniform during the whole year. The labor of churning whole milk is, of course, much greater than if the cream alone were employed, but the increased yield and unvarying quality of the butter more than compensate for the extra expenditure of labor.

The proper temperature of the milk or cream is a point of great practical importance. If the fluid be too warm or too cold the buttery particles will only by great trouble be made to cohere; and the quality of the butter is almost certain to be inferior. When the whole milk is operated on, the temperature should be from 55 to 60 degs. of Fahrenheit's thermometer; and if cream be employed the temperature should never exceed 55 degs. nor be lower than 50 degs. Hence it follows that in summer the dairy should be kept cooler, and in winter warmer, than the atmosphere. The temperature of milk is raised or lowered as may be found necessary, by the addition of hot or cold water—in performing which operations properly, a good thermometer is indispensable; one should always be kept in the dairy, and should be so constructed as to admit of being plunged into the milk. In some dairies the water, instead of being mixed with the milk, is put into a tub in which the churn is placed. There is a good kind of churn, which consists of two cylinders, the one within the other—the interval between them being intended for the reception of hot or cold water. The influence of temperature upon the production of butter has been placed beyond all doubt by numerous carefully-conducted experiments. Mr. Horsfall, a celebrated dairy farmer, in discussing this question, sums up as follows:—" By a series of carefully-conducted experiments at varying temperatures, I am of opinion that a correct scale of the comparative yield of butter at different temperatures might be arrived at; as thus: From a very low degree of temperature little or no butter; from a temperature of about 38 degs., 16 oz. from 16 quarts of milk; ditto, 45 degs., 21 oz. from 16 quarts of milk; ditto, 55 degs., 26 to 27 oz. from 16 quarts of milk." This is a higher yield of butter than, I suspect, most dairymen get: but Mr. Horsfall's cows being of the best kind for milking, and well fed, the milk is, of course, rich in butter; and his experiments prove that even the richest milk will not throw up its butter unless at a certain temperature.

In the churning of cream the motion should be slow at first until the cream is thoroughly broken up. In churning milk the agitation should neither be violent nor irregular; about 40 or 50 motions of the plunger or board per minute will be sufficient. In steam-worked churns the motion is often excessively rapid, and the separation of the butter is effected in a few minutes; but the article obtained in this hasty way very quickly becomes rancid, and must be disposed of at once. An hour's churning of sour cream appears in general to produce good butter. Sweet cream and whole milk require a longer period —the latter about 3 hours—but in any case prolonged churning is certain, by incorporating cheesy matter with the butter, to produce an inferior article.

Sweet milk becomes sour, evolves a considerable quantity of gas during churning, and its temperature ascends four or five degrees. Oxygen is unquestionably absorbed, and it is probable that a portion of the sugar of milk is converted into acid products.

I have already stated that even the most carefully prepared butter contains a small proportion of casein and sugar of milk. This casein is the good genius of the cheese-maker, but the evil genius of the butter manufacturer. How? In this way:— When butter containing a notable proportion of casein and sugar of milk is exposed to the air, the following changes take place: the casein passes into a state of fermentation, and acting upon the sugar of milk, converts it, firstly into the bad-flavored lactic acid, and secondly into the bad odorous butyric, capric, and caproic acids. The first of these compounds in a state of purity emits an odor resembling a mixture of vinegar and rancid butter; the second possesses an odor resembling that of a goat —hence the name *capric;* the third has an odor like that of perspiration. In addition to these acids, there is another simultaneously generated—the caprylic, but it does not unpleasantly affect the olfactory nerve. The casein also injuriously affects the fatty constituents of the butter; under its influence they absorb oxygen from the air, and become converted into strong-smelling compounds. The washing of butter is intended to free it from the casein and unaltered cream, and the more perfectly it is freed from those impurities the better will be its flavor, and the longer it will remain without becoming rancid. Some people believe that too much water injures the quality and lessens the quantity of butter. It cannot do the former, because the essential constituents of butter are totally insoluble in water; it may do the latter, but, if it do, so much the better, because the loss of weight represents the amount of impurities—milk, sugar of milk, &c.—removed.

I have already remarked that butter is so susceptible of taint that even a perspiring hand is sufficient to spoil it; naturally cool hands should alone be allowed to come in contact with this delicate commodity, and the hands should be made thoroughly clean by repeated washings with warm water and oatmeal—the use of soap in the lavatory of the dairymaid being highly objectionable. Wooden spades are now being commonly made use of in manipulating the butter, and there is no good reason why they should not come into universal use.

The yield of butter per cow is subject to great variation. Some breeds of the animal are remarkable as milkers; such, for instance, as the Alderneys and Kerrys—indeed, I may say all the small varieties of the bovine race. There are instances of cows yielding upwards of twenty pounds of butter per week, but these are extraordinary cases. In Holland a good cow will produce, during the summer months, more than 180 lbs. of butter. In these countries I think the average annual yield of a cow is not more than 170 lbs. It sometimes happens that cows yield a large quantity of milk and a small amount of butter, but it far more frequently occurs that the cow which gives most milk also yields most butter.

An estimate of the amount of butter contained in milk may be made by determining the amount of cream. This may be effected by means of an instrument termed a *lactometer*, which is simply a glass tube about five inches long, and graduated into a hundred parts. The specimen to be examined is poured into this tube up to zero or o, and allowed to stand for twelve hours in summer and sixteen or eighteen in winter. At the end of that time the cream will have risen to the top, and its per-centage may be easily seen. In good milk the cream will generally extend 11 to 15 degrees down from o. This instrument, although very useful, is not reliable in every case, especially in detecting the adulteration of milk.

I have already stated that the complete separation of the butter from the other constituents of the milk is never accomplished in the dairy. Now although the proportion of curd in the butter is very small—rarely more than two per cent., and often not a fourth of one per cent.—yet it is more than sufficient, under a certain condition, to cause the butter to become speedily rancid. That condition is simply contact with the air. If the curd, before it becomes dry and firm, is subjected to the influence of the air, it rapidly passes into a state of fermentation, which is very soon communicated to the fatty and saccharine constituents of the butter (substances not spontaneously liable to sudden changes in composition) and those peculiar compounds—such, for example, as butyric and capric acids, are generated, which confer upon rancid butter its characteristic and very disagreeable odor and flavor. The fermentation of the curd is prevented by incorporating common salt with the butter, and by preventing, so far as possible, the access of air to the vessels in which the article is placed. If fresh butter be placed in water—which apparently protects it from the influence of the air—it will soon become rancid. The reason of this is, that water always contains air, which differs in composition, though derived, from the atmosphere, by being very rich in oxygen. Now, it is precisely this oxygen which effects those undesirable changes in the casein, or curd, to which I have so repeatedly referred; hence its presence in a concentrated state in water causes that fluid to produce an injurious effect on the butter placed in it. A saturated solution of salt contains very little air, and, so long as the curd is immersed therein, it undergoes no change. The salt, too, acts as a decided preservative; for although it was long considered to be capable of preserving animal matters, merely by virtue of its property of absorbing water from them (the presence of water being a condition in the decomposition of organic matter), it has lately been shown to possess very antiseptic properties.

The mixing of the salt with the butter is effected in the following manner:—The butter, after being well washed, in order to free it from the butter-milk, is spread out in a tub, and the salt shaken over it; the butter is then turned over on the salt by the lower part of the palm of the hand, and rubbed down until a uniform mixture is attained. A good plan in salting is to mix in only one half of the quantity of salt, make up the butter in lumps, and set them aside until the following day; a quantity of milk is certain to exude, which is to be poured off, and then the rest of the salt may be incorporated with the butter.

According to butter-makers, the quality of the article is greatly dependent on the quality of the salt used in preserving it. I think there is a good deal of truth in this belief, and I therefore recommend that only the very best and *driest* salt should be used in the dairy. Common salt is essentially composed of the substance termed by chemists chloride of sodium, but it often contains other saline matters (chloride of magnesium, &c.), some of which have a tendency to absorb moisture from the air, and to dissolve in the water so obtained. These salts are termed *deliquescent*, from the Latin *deliquere*, to melt down. When, therefore, common salt becomes damp by mere exposure to the air, it is to be inferred that it contains impurities which, as they possess a very bitter taste, would, if mixed with butter, confer a bad flavor upon it. The impurities of salt may be almost completely removed by placing about a stone weight of it in any convenient vessel, pouring over it a quart of boiling water, and mixing thoroughly the fluid and solid. In an hour or two the whole is to be thrown upon a filter made of calico, when the water will pass through the filter, carrying with it all the impurities, and the purified salt, in fine crystals, will remain upon the filter. The solution need not be thrown away: boiled down to dryness it may be given as salt to cattle; or, if added in solution to the dung-heap, it will augment the fertilising power of that manure.

The proportion of salt used in preserving butter varies greatly. When the butter is intended for immediate use, I believe a quarter of an ounce of salt to the pound is quite sufficient; but when designed for the market, about half an ounce of salt to the pound of butter will be sufficient. Irish butter at one time commanded the highest price in the home and foreign markets, but latterly it has fallen greatly in public estimation; indeed, at the present moment the price of Irish butter at London is nearly twenty shillings per cwt. under that of the Dutch article. It is really painful to be obliged to admit that the Irish farmer is solely to blame for this remarkable depreciation in the value of one of our best agricultural staples. In a word, by the stupid (and *recent*) practice of putting into butter four times the quantity of salt necessary to its preservation, the Irish dairy farmers—or at least the great majority of them—have completely ruined the reputation of Irish butter in those very markets in which, at one time, the Cork brand on a firkin was sufficient to dispose of its contents at the very highest price. It is a great mistake to think that the greater the quantity of salt which can be incorporated with the butter, the greater will be the profit to the producer. No doubt, every pound of salt sold as a constituent of butter realises a profit of two thousand per cent.; but then the addition of every pound of that substance, after a certain quantity, to the cwt. of butter depreciates the value of the latter to such an extent as to far more than neutralise the gain on the sale of salt at the price of butter. In the county of Carlow, less salt is used in preserving butter than is the case in the county of Cork and the adjacent counties; the price, therefore, which the Carlow commodity commands in the London market is higher than that of the Cork butter: but in every part of Ireland the proportion of salt added to the butter is excessive.

The results of the analyses of butter supplied to the London market, made by the *Lancet* Analytical Commission, showed that the proportion of salt varied from 0.30 to 8.24 per cent. The largest proportion of salt found in fresh butter was 2.21 and the least 0.30. In salt butter the highest proportion of salt was 8.24 and the lowest 1.53. The butter which contained most salt was also generally largely adulterated with water. Indeed, in several samples the amount of this constituent reached so high as nearly 30 per cent. Nothing is easier than the incorporation of water with salt butter. The butter is melted, and whilst cooling the salt and water are added, and the mixture kept constantly stirred until quite cold. In this way nearly 50 per cent. of water may be added to butter; but of course the quality of the article will be of the very worst kind.

A correspondent of the *Lancet* states that, on awakening about three o'clock in the morning at the house in which he was lodging, he perceived a light below the door of his room; and apprehending a fire, he hurried down stairs, and was not a little surprised to discover the whole family engaged in manipulating butter. He was informed in a jocose way that they were making Epping butter! For this purpose they used inferior Irish butter, which, by repeated washings, was freed from its excessive amount of salt; after which it was frequently bathed in sweet milk, the addition of a little sugar being the concluding stroke in the process. This "sweet fresh butter from Epping" was sold at a profit of 100 per cent. Our dairy farmers might take a hint from this anecdote. Does it not prove that the mere removal of the salt added to Irish butter doubles the value of the article?

It is as necessary to pay attention to the packing of butter as it is to its salting. If old firkins be employed, great care should be taken in cleaning them, and if the staves be loose, the firkins should be steeped in hot water, in order to cause the wood to swell, and thereby to bring the edges of the staves into close contact. New firkins often communicate a disagreeable odour to the butter. In order to guard against this, it is the practice in many parts to fill the firkins with very moist garden mould, which, after the lapse of a few days, is thrown out, and the firkin thoroughly scrubbed with hot water, rinsed with the same fluid in a cold state, and finally rubbed with salt, just before being used.

In packing the butter, the chief object to be kept in view is the exclusion of air. In order to accomplish this, the lumps of butter should be pressed firmly together, and also against the bottom and sides of the vessel. When the products of several churnings are placed in the same firkin, the surface of each churning should be furrowed, so that the next layer may be mixed with it. A firkin should never be filled in a single operation. About six inches of butter of each churning will be quite sufficient, and in a large dairy two or more firkins can be gradually but simultaneously filled. I strongly recommend the removal of the pickle jar from the dairy. When the layers of butter have been carried up to within an inch or so of the top of the firkin, the space between the surface of the butter and the edge of the vessel should be filled with fine dry salt, instead of pickle. A common mistake made is the holding over for too long a time of the butter: the sooner this article can be disposed of the better, for *it never improves by age.*

PART V.

ON THE COMPOSITION AND NUTRITIVE VALUE OF VEGETABLE FOODS.

SECTION I.

THE MONEY VALUE OF FOOD SUBSTANCES.

THE flesh-forming principles of food are, as I have already stated, almost identical with the principal nitrogenous constituents of animals. Unlike the non-plastic substances, they are convertible into each other with little, if any, loss either of matter or of force. Not many years since it was the fashion to estimate the nutritive value of a food-substance by its proportion of nitrogen; but this method—not yet quite abandoned—was based on erroneous views, and yielded results very far from the truth. No doubt all the more concentrated and valuable kinds of food are rich in nitrogenous principles; but there are other varieties, the nutritive value of which is very low, and yet their proportion of nitrogen is very high: This point requires explanation. Both the plastic and the non-plastic materials of food exist in two distinct states—in one of which they are easily digestible, and in the other either altogether unassimilable or so nearly so as to be almost useless. Thus, for example, the cellular tissue of plants, when newly formed, is to a great extent digestible, whilst the old woody fibre is nearly, if not quite, incapable of assimilation. Gelatine, which in raw bones is easily digested in the stomachs of the carnivora, loses a large proportion of its nutritive value on being subjected to the action of steam. Again, a portion of the nitrogen of young succulent plants is in a form not sufficiently organic to admit of its being assimilated to the animal body. But, independently of these strong objections to the method of estimating the nutritive value of food by its per-centage of flesh-formers, there are many other reasons which as clearly prove the fallacy of this rule. If we were, for instance, to estimate the value of albumen according to the tables of food equivalents which were constructed some years ago by Boussingault and other chemists, we would find one pound weight of it to be equivalent to four pounds weight of oil-cake, or to twelve pounds weight of hay; yet, it is a fact that a horse would speedily die if confined to a purely albuminous diet, whereas hay is capable of supporting the animal's life for an indefinite period.

It is clear, then, from what I have stated, that neither the amount of flesh-formers, nor of fat-formers, contained in a given quantity of a substance is a measure of its nutritive value; nevertheless it would be incorrect to infer from this that the numerous analyses of feeding substances which have been made are valueless. On the contrary, I am disposed to believe that the composition of these substances, when correctly stated by the chemist, enables the physiologist to determine pretty accurately their relative alimentary value. Theory is certainly against the assumption that food is valuable in proportion to its content of nitrogen; nor has practice less strongly disproved its truth. An illustration drawn from the nutrition of plants will make this matter more apparent. Every intelligent agriculturist knows that guano contains nitrogen and phosphoric acid; both substances are indispensable to the development of plants, and therefore it would be incorrect to estimate the manurial value of the guano in proportion to the quantity of nitrogen it was capable of yielding. If the value of manures were determined only by their per-centage of nitrogen—a mode by which certain chemists still estimate the nutritive value of food—then woollen rags would be worth more than bones, and bones would be more valuable than superphosphate of lime. The truth is, that the analysis of feeding stuffs and manures is sometimes of little value if the condition in which the constituents of these substances exist be undetermined. For example, the analysis of one manure may show it to contain 40 per cent. of phosphate of lime, and three per cent. of ammonia, whilst, according to analysis, another fertiliser may include 20 per cent. of phosphate of lime, and two per cent. of ammonia. Viewed by this light solely, the first manure would be considered the more valuable of the two, whereas it might, in reality, be very much inferior. If the phosphate of lime in the manure, containing 40 per cent. of that body, were derived from coprolites or apatite, and its ammonia from horns, the former would be worth little or nothing, and the latter, by reason of its exceedingly slow evolution from the horns, would possess a very low value. If, on the contrary, the phosphate of lime, in the manure comparatively poor in phosphate, were a constituent of bones, and its ammonia ready formed (say as sulphate of ammonia), then, its value, both commercial and manurial, would be far greater than the other.

In estimating the money value of an article of food, we should omit such considerations as the relative adjustment of its flesh-formers and fat-formers, and its suitability to particular kinds of animals, as well as to animals in a certain stage of development. The manure supplied to plants contains several elements indispensable to vegetable nutrition; and, although the agriculturist most commonly purchases all these elements combined in the one article, still he frequently buys each ingredient separately. Ammonia is one of these principles, and, whether it be bought *per se*, or as a constituent of a compound manure, the price it commands is invariable. This principle should prevail in the purchase of food: each constituent of which should have a certain value placed upon it; and the sums of all the values of the constituents would then be the value of the article of food taken as a whole. There are, no doubt, practical difficulties in the way which prevent this method of valuation from giving more than approximately correct results; but are there not precisely similar difficulties in the way of the correct estimation of the value of a manure according to its analysis? There are several constituents of food, the money value of which is easily determinable: these are sugar, starch, and fat. No matter what substance they are found in, the nutritive value of each varies only within very narrow limits. The value of cellulose and woody fibre is not so easily ascertained, as it varies with the age and nature of the vegetable structure in which these principles occur. There is little doubt but that the cellulose and fibre of young grass, clover, and other succulent plants, are, for the most part, digestible; and we should not be far astray if we were to assume that four pounds weight of soft fibre and cellulose are equivalent to three pounds weight of starch. As to old hard fibre, we are not in a position to say whether or not it possesses any nutrimental value worth taking into account. The estimation of the value of the flesh-forming materials is far more difficult than that of sugar, starch, pectine compounds, and fat. The nitrogenous constituents of food must be in a highly elaborated state before they are capable of being assimilated. In seeds—in which vegetable substances attain their highest degree of development—they probably exist in the most digestible form, whilst much of the nitrogen found in the stems and leaves of succulent plants, is either in a purely mineral state, or in so low a degree of elaboration as to be unavailable for the purpose of nutrition. But even plastic materials, in a high degree of organisation, present many points of difference, which greatly affect their relative alimental value; for example, many of them are naturally associated with substances possessing a disagreeable flavor: and as their separation from these substances is often practically impossible, the animal that consumes both will not assimilate the plastic matters so well as if they were endowed with a pleasant flavor. In seeds and other perfectly matured vegetable structures, the flesh-formers may exist in different degrees of availability. The nitrogen of the *testa*, or covering of the seeds, will hardly be so assimilable as that which exists in their cotyledons. The solubility of the flesh-formers—provided they be highly elaborated—is a very good criterion of their nutritive power. In linseed the muscle-forming substances are more soluble than in linseed-cake—the heat which is generally employed in the extraction of oil from linseed rendering the plastic materials of the resultant *cake* less soluble, and diminishing thereby their digestibility, as practice has proved.

From the considerations which I have now entered into, it is obvious that the chemical analysis of food substances as generally performed, though of great utility, does not afford strictly accurate information as to their commercial value, and still less reliable in relation to their nutritive power. At the same time, they as clearly establish the feasibility of analyses being *made* whereby the money value of feeding-stuffs may be estimated with tolerable exactitude. Let the chemist determine the presence and relative amounts of the ingredients of food-substances, and—if it be possible so to do with a degree of exactness that would render the results useful—place on each a money value. This done, let the physiologist and the feeder combine the food in such proportions as they may find best adapted to the nature, age, and condition of the animal to be fed.

It is to be regretted that the market price of feeding stuffs is not, in consequence of our defective knowledge, strictly determined by their nutritive value, for if such were the case, the feeder would merely have to adapt each to the nature and condition of his stock. Even amongst practical men there prevails, unfortunately, great diversity of opinion as to the relative nutritive value of the greater number of food substances; and I am quite certain that many of these command higher prices than others which in no respect are inferior. It would lead me too far from my immediate subject were I to enter minutely into the consideration of such questions as —whether an acre of grass yields more or less nutriment than an acre of turnips? I shall merely describe the composition and properties of grass and of turnips, and of the various other important food substances, and compare their nutritive power, so far as comparisons are admissible; but I shall say but little on the subject of the various economic and other conditions which affect the production of forage plants. When I shall have described the chemical nature and physical condition of the various articles of food, and the results of actual feeding experiments made with them, the feeder will then be in a position to determine which are the most economical to produce or to purchase.

SECTION II.

PROXIMATE CONSTITUENTS OF VEGETABLES.

The saccharine, or amylaceous substances constitute the most abundant of the proximate constituents of plants. They are composed of carbon, hydrogen, and oxygen. I shall briefly describe the more important members of this group of substances, namely, starch, sugar, inulin, gum, pectin, and cellulose.

Starch, or *fecula*, occurs largely in dicotyledonous seeds, peas, &c., and still more abundantly in certain monocotyledonous seeds, such as wheat and barley. It constitutes the great bulk of many tubers and roots—for example, the potato and tapioca. It consists of flattened ovate granules, which vary in size according to the plant. In the beetroot they are $\frac{1}{3500}$ of an inch in diameter, whilst in *tous les mois* they are nearly $\frac{1}{300}$ of an inch in diameter. Most of the starch granules are marked by a series of concentric rings. Starch is heavier than water, and is insoluble in that fluid when cold; neither is it dissolved by alcohol or ether. When heated in water having a temperature of at least 140° Fahrenheit, it increases greatly in volume, and acquires a gelatinous consistence. When the water is allowed to cool, a portion of the starch becomes insoluble, whilst another portion remains in solution; the latter form of starch is sometimes termed *amidin*, from the French word for starch, *amidon*. When dry starch is heated to 400° Fahr., it is converted, without any change in its composition, into a soluble gum-like substance, termed *dextrin*, or British gum. On being boiled in diluted sulphuric acid it is converted into a kind of sugar; and the same effect is produced by fermentation—for example, in the germination of seeds. Fresh rice contains 82, wheat 60, and potatoes 20 per cent. of starch. This substance constitutes a nutritious and easily digestible food, but alone cannot support life. Arrowroot is only a pure form of starch.

Sugar occurs less abundantly in plants than starch. There are several varieties of this substance, of which the kinds termed cane sugar (*sucrose*) and grape sugar (*glucose*), are only of importance to agriculturists. The former enters largely into the composition of the sugar-cane, the beetroot, the sugar-maple, the sorgho grass, pumpkins, carrots, and a great variety of other plants. Grape sugar is found in fruits, especially when dried—raisins and figs—in malted corn, and in honey. In the sugar-cane there is 18 per cent., and in the beetroot 10 per cent. of sugar.

Cane sugar, when pure, consists of minute transparent crystals. It is $1\frac{6}{10}$ heavier than water, and is soluble in one-third of its weight of that fluid. By long-continued boiling in water it is changed into uncrystallizable sugar, or treacle, by which its flavor is altered, but its sweetening power increased.

Grape sugar crystallizes in very small cubes, of inferior color as compared with cane sugar crystals. It dissolves in its own weight of water, being three times less soluble than sucrose. In sweetening power one part of cane sugar is equal to $2\frac{1}{2}$ parts of grape sugar; but there is probably little if any difference, between the nutritive power of the two substances.

Inulin is a substance somewhat resembling starch. It does

not occur in large quantities. It is met with in the roots of the dandelion, chicory, and many other plants.

Gum is an abundant constituent of plants. The kind termed gum arabic, so largely employed in the arts, is a very pure variety of this substance. Common gums are said to be essentially composed of a very weak acid—*gummic*, or *arabic* acid—united with lime and potash. The solution of gum is very slightly acid, and has a mucilaginous, ropy consistence: it is almost tasteless. *Mucilage*, or *bassorin*, is simply a modified form of gum, which, though insoluble in water, forms a gelatinous mixture with that fluid. It exudes from certain trees—the cherry for example—and exists largely in linseed and other seeds. Gums are nutritious foods, but it is probable that they are not equal in alimental power to equal weights of starch or sugar.

Vegetable jelly, or *pectin*, is almost universally diffused throughout the vegetable kingdom. It is owing to its presence that the juices of many fruits and roots possess the property of gelatinizing. It is soluble in water, but prolonged boiling destroys its viscous property. *Pectose* is a modification of pectin; it is insoluble in water. According to Fremy, the hardness of green fruits is due to the presence of pectose; which is also found in the cellular tissue of turnips, carrots, and various other roots.

Cellulose is a fibrous or cellular tissue, allied in composition to starch. It is the most abundant constituent of plants, and forms the very ground-work of the vegetable mechanism. Linen, cotton, and the pith of the elder and other trees are nearly pure forms of cellulose. Ligneous, or woody tissue (*lignin*) is indurated cellulose, hardened by age. It is almost identical in composition with cellulose. Pure cellulose is white, colorless, tasteless, insoluble in water, oil, alcohol, or ether. It is heavier than water. Sulphuric acid is capable of converting it into grape, or starch sugar. In its fresh and succulent state cellulose is digestible and nutritious; but in the form of ligneous tissue it opposes a very great resistance to the action of the digestive fluids. Digestible cellulose is probably equal in nutritive power to starch.

Oils and fats occur abundantly in vegetables, more particularly in their seeds. In the seeds of many cruciferous plants the proportion of fat and oil exceeds 35 per cent. The oils and fats termed *fixed* are those which possess the greatest interest to agriculturists; the *volatile oils* being those which confer on certain plants their fragrant odour. There are a great variety of vegetable oils, but the proximate constituents of most of them are chiefly *stearin, margarin, olein,* and *palmitin.*

Stearin is a white crystalline substance, sparingly soluble in alcohol and ether, but insoluble in water. There are two or three modifications of this substance, but they do not essentially differ from each other. The melting point varies from $130°$ to $160°$ Fahr. Stearin is the most abundant of the fats.

Margarin presents the appearance of pearly scales. It is the solid fat present in olive oil, and it is also met with in a great variety of fats and oils. It melts at $116°$ Fahr.

Olein is the fluid constituent of oils and fatty substances. It resists an extreme degree of cold, without solidifying. There are several modifications of this body—the olein of olive oil being somewhat different from that of castor oil; the olein of linseed is sometimes termed *linolien.*

Palmitin.—This fat occurs in many plants, but as it makes up the great bulk of palm oil, it has been termed palmitin. It is white, and may be obtained in feathery-like masses. Its melting point varies from $114°$ to $145°$, there being, according to Duffy, three modifications of this substance.

The fats and oils are lighter than water. They contain far more carbon and hydrogen, and less oxygen, than are found in the sugars and starches. They all consist of acids (stearic, palmitic, &c.) united with glycerine. On being boiled with potash or soda, the latter take the place of the glycerine, which is set free, and a *soap* is produced. The fatty acids strongly resemble the fats. In nutritive power, one part of fat is equal to $2\frac{1}{2}$ parts of starch or sugar.

The Albuminous substances contain, in addition to the elements found in starch, nitrogen, sulphur, and phosphorus. *Albumen*, *fibrin*, and *legumin* constitute the three important members of the "Nitrogenous" constituents of plants.

Albumen is an uncrystallizable substance. It is soluble in water, unless when heated to 140 deg. Fahr., at which temperature it coagulates, *i.e.*, becomes solid and insoluble. The *gluten* of wheat is composed chiefly of albumen, and of bodies closely allied to that substance.

Fibrin, when dried, is a hard, horny, yellow, solid body. It contains a little more oxygen than is found in albumen. This substance is best known as a constituent of animals, and it does not appear to be abundant in plants. The portion of the gluten of wheat-flour, which is insoluble in boiling alcohol, is considered by Liebig and Dumas to be coagulated fibrin.

In the seeds of leguminous and a few other kinds of plants large quantities of a substance termed *legumin* are found. It resembles the casein, or cheesy ingredient of milk; indeed, some chemists consider it to be identical in composition with that substance. When pure, it is pearly white, insoluble in boiling water, but soluble in cold water and in vinegar. The saline matters found in plants are always associated with the albuminous bodies; the latter, therefore, form the bones as well as the muscles of animals.

A great many substances are found in plants, such as wax, mannite, "extractive matter," citric, malic, and other acids, of the nutritive value of which very little is known. The substances described in this section constitute, however, at least 95 per cent. of the weight of the vegetable matters used as food by live stock.

SECTION III.

GREEN FOOD.

The Grasses.—More than one-half the area of Great Britain and Ireland is under pasture; the grasses, therefore, constitute the most important and abundant food used by live stock.

The composition of the natural and artificial grasses is greatly influenced by the nature of the soil on which they are grown, and by the climatic conditions under which they are developed. Many of them are almost worthless, whilst others possess a high nutritive value. Amongst the most useful natural grasses many be enumerated Italian rye-grass, Meadow barley, Annual Meadow-grass, Crested dogstail-grass, Cocksfoot-grass, Timothy or Meadow catstail-grass, and Sweet vernal-grass. Amongst grasses of medium quality I may mention common Oatlike-grass, Meadow foxtail grass, Smooth and rough stalked Meadow-grass, and Waterwhorl-grass. There are very many grasses which are almost completely innutritious, and which ought, under no circumstances, to be tolerated, although too often they make up the great bulk of the herbage of badly-managed meadows and pastures. Such grasses are, the Meadow soft-grass, Creeping soft-grass, False brome-grass, and Upright brome-grass. The rough-stalked Meadow-grass, though spoken favorably of by some farmers, is hardly worthy of cultivation, and the same may be said of many of the grasses which have a place in our meadows and pastures. (See "Analyses of Natural Grasses in a Fresh State, by Dr. Voelcker," on next page.)

The *Schræder brome* is a perennial lately introduced into France. It is described as an exceedingly valuable forage crop, and one which is admirably adapted for the feeding of dairy cows. It would be desirable to give it a trial in these countries. The composition (which is very peculiar) of this plant is stated to be as follows, when dry:—

ANALYSIS OF SCHRÆDER BROME HAY.

Water	16·281
Nitrogenous matters	23·443
Fat	3·338
Starch gum, &c.	22·549
Cellulose (fibre)	19·843
Ashes	14·546
Total	100·000

ANALYSES OF NATURAL GRASSES IN A FRESH STATE, BY DR. VOELCKER.

	Water.	Albuminous or Flesh-forming Principles.	Fatty Matters.	Respiratory Principles: Starch, Gum, Sugar.	Woody Fibre.	Mineral Matter or Ash.	Date of Collection.
Anthoxanthum odoratum — Sweet-scented vernal grass	80·35	2·00	·67	8·54	7·15	1·24	May 25
Alopecurus pratensis—Meadow foxtail grass.	80·20	2·44	·52	8·59	6·70	1·55	June 1
Arrhenatherum avenaceum—Common oat-like grass	72·65	3·54	·87	11·21	9·37	2·36	July 17
Avena flavescens—Yellow oat-like grass ...	60·40	2·96	1·04	18·66	14·22	2·72	June 29
,, pubescens—Downy oat-grass ...	61·50	3·07	·92	19·16	13·34	2·01	July 11
Briza media—Common quaking grass ...	51·85	2·93	1·45	22·60	17·00	4·17	June 29
Bromus erectus—Upright brome grass ...	59·57	3·78	1·35	33·19		2·11	June 23
,, mollis—Soft brome grass	76·62	4·05	·47	9·04	8·46	1·36	May 8
Cynosurus cristatus—Crested dogstail grass ..	62·73	4·13	1·32	19·64	9·80	2·38	June 21
Dactylus glomerata—Cocksfoot grass ...	70·00	4·06	·94	13·30	10·11	1·54	,, 13
Ditto, seeds ripe	52·57	10·93	·74	12·61	20·54	2·61	July 19
Festuca duriuscula—Hard fescue grass...	69·33	3·70	1·02	12·46	11·83	1·66	June 13
Holcus lanatus—Soft meadow grass ...	69·70	3·49	1·02	11·92	11·94	1·93	,, 29
Hordeum pratense—Meadow barley ...	58·85	4·59	·94	20·05	13·03	2·54	July 11
Lolium perenne—Darnel grass	71·43	3·37	·91	12·08	10·06	2·15	June 8
,, italicum—Italian rye-grass ...	75·61	2·45	·80	14·11	4·82	2·21	,, ,,
Phleum pratense—Meadow catstail grass ...	57·21	4·86	1·50	22·85	11·32	2·26	,, 13
Poa annua—Annual meadow grass ...	79·14	2·47	·71	10·79	6·30	·59	May 28
,, pratensis—Smooth-stalked meadow grass	67·14	3·41	·86	14·15	12·49	1·95	June 11
,, trivialis—Rough-stalked ditto ...	73·60	2·58	·97	10·54	10·11	2·20	,, 18
Grass from water meadow	87·58	3·22	·81	3·98	3·13	1·28	April 30
Ditto, second crop	74·53	2·78	·52	11·17	8·76	2·24	June 26
Annual rye-grass	69·00	2·96	·69	12·89	12·47	1·99	,, 8

Most of the grasses here mentioned were analysed when in flower.

Tussac Grass (*Dactylis cæspitus*) is recommended as an excellent plant to grow on very poor, wet, or mossy soils.* It is an evergreen grass, somewhat resembling coltsfoot. It is relished by cattle.

ANALYSIS OF TUSSAC GRASS BY JOHNSTONE.

	Lower part.	Upper part.
Water	86·09	75·17
Flesh-formers	2·47	4·79
Sugar, gum, &c.	4·62	6·81
Woody fibre (with a little albumen)	5·68	11·86
Ash	1·14	1·37
Total	100·00	100·00

The "artificial grasses" embrace the clovers, vetches, lucerne, and a few other plants, some of which are seldom cultivated.

ANALYSES OF DIFFERENT KINDS OF CLOVER, BY DR. ANDERSON.

	Per-centage in the Fresh Clover.				Per-centage in Dry Clover.	
	Water.	Dry Substances.	Ash.	Nitrogenised Substances.	Ash.	Nitrogenised Matters.
Red clover—Trifolium pratense:						
1. From English seed...	85·30	14·70	1·30	2·31	8·90	15·87
2. From German seed (from the Rhine) ...	81·68	18·32	1·49	2·81	8·15	15·50
3. From French seed ...	83·51	16·49	1·95	2·25	11·82	13·56
4. From American seed	79·98	21·02	1·58	2·87	8·05	...
5. From Dutch seed ...	...	...	...	...	8·82	12·43
Cowgrass — Trifolium medium:†						
Variety, Duke of Norfolk	77·39	22·61	2·73	2·25	12·09	10·19
,, common	81·76	18·24	1·92	3·19	10·53	14·37
Crimson clover, Trifolium incarnatum:						
From French seed ...	82·56	17·44	1·88	3·25	10·81	18·56
Yellow clover — Medicago lupulina:						
From English seed ...	77·38	22·62	2·02	3·50	8·95	15·44
From French seed ...	78·60	21·40	1·75	2·94	8·18	13·69

* See Transactions of Highland and Agricultural Society of Scotland for 1852.
† Zig-zag clover, or Marl grass? Cowgrass is *Trifolium pratense perenne*.

Clover is very rich in flesh-forming and heat-producing substances. There are several varieties of this plant, of which the Alsike Clover appears to be the most valuable, as it contains a high proportion of organic matter and gives the largest acreable produce. The nature of the soil influences, to a great extent, the composition of this plant: this no doubt accounts for the somewhat discrepant result of the analyses of it made by Way, Voelcker, and Anderson.

The composition of the Vetch, Sainfoin, and Lucerne, resembles very closely that of the Clover: indeed, it appears to me that all these leguminous plants are nearly equally valuable as green forage, but that the best adapted for hay is the Clover. In the following table the composition of these plants is shown:—

ANALYSES OF CLOVER, BY DR. VOELCKER.

	I. Red Clover.	II. White Clover.	III. Yellow Clover.	IV. Alsike Clover.	V. Bokhara Clover.
Water	80·64	83·65	77·57	76·67	81·30
Soluble in Water—					
a. Organic substances	6·35	4·98	8·26	4·91	6·80
b. Inorganic substances	1·55	1·13	1·40	1·33	1·54
Insoluble in water—					
a. Impure vegetable fibre	11·04	9·80	12·17	16·36	10·01
b. Inorganic matters (ash)	0·42	0·44	0·60	0·73	0·35
	100·00	100·00	100·00	100·00	100·00

ANALYSES OF LUCERNE, SAINFOIN, AND VETCH.

	I. Lucerne.	II. Sainfoin.	III. Vetch.
Water	73·41	77·32	82·16
Soluble in Water—			
a. Organic substances	9·43	8·00	6·07
b. Inorganic substances	2·33	1·20	1·07
Insoluble in water—			
a. Impure vegetable fibre	14·08	12·95	10·23
b. Inorganic matters (ash)	0·75	0·53	0·47
	100·00	100·00	100·00

The artificial grasses are, on the whole, more nutritious than the natural grasses; but I should explain that the analyses of the natural grasses which I have quoted refer to those plants in what may be almost termed their wild state : under the influence of good cultivation—when irrigated or top-dressed with abundance of appropriate manure—their analyses would indicate a higher nutritive value. The grasses, and more especially the so-called artificial grasses, are more nutritious and digestible when young. In old clover the proportion of insoluble woody fibre is often so considerable as to greatly detract from the alimental value of the plant.

The *Lentils*, the *Birdsfoot*, the *Trefoil*, and the *Melilot* are leguminous plants which occasionally are found as constituents of forage crops. Lentils are extensively cultivated on the Continent, and are the only kind of these plants the chemistry of which has been at all studied. The straw contains 7 per cent. of flesh-formers.

The Yellow Lupine is cultivated rather extensively in Germany, France, and Belgium, partly for feeding purposes, partly to furnish a green manure. Its seeds constitute a nutritious article of food for man, and its stems and leaves are given to cattle. An attempt was made a few years ago to introduce its cultivation, as a field crop, into England, and very satisfactory results attended the first trials made with it. Mr. Kimber, who has cultivated this crop, states that it is likely to prove valuable on light sandy soils, where the ordinary green fodder crops are not easily cultivated. The produce per acre obtained in Mr. Kimber's trial was about nineteen tons. Cattle and sheep relish the Yellow Lupine, but according to Mr. Kimber, pigs reject it. Professor Voelcker examined this plant, and found that it resembled in composition the ordinary artificial grasses, except in one respect, namely, a remarkable deficiency in sugar. Altogether, it is not so rich in nutriment as any of the commonly cultivated leguminous plants; but as it can be cultivated on a very poor soil, and gives a good return, it is probable that the Yellow Lupine will yet become a common crop in Britain. The following table exhibits the results of Dr. Voelcker's analysis.

COMPOSITION OF YELLOW LUPINES (CUT DOWN IN A GREEN STATE).

	In natural state.	Dried at 212° F.
Water	89·20	
Oil	·37	3·42
*Soluble albuminous compounds	1·37	12·68
Soluble mineral (saline) substances	·61	5·64
†Insoluble albuminous compounds	1·01	9·35
Sugar, gum, bitter extractive matter, and digestible fibre	3·96	36·68
Indigestible woody fibre (cellulose)	3·29	30·48
Insoluble mineral matters	·19	1·75
	100·00	100·00
* Containing nitrogen	·22	2·03
† Containing nitrogen	·16	1·48

Rib grass plantain (Plantago lanceolata) is one of those plants, the value of which for forage purposes is questionable. Many persons believe it to be a useful food. Its composition, which looks favorable, is as follows :—

Water	84·78
Albuminous matters	2·18
Fatty matters	0·56
Starch, gum, &c.	6·08
Woody fibre	5·10
Mineral matter	1·30

The grasses, natural and artificial, are occasionally affected by a formidable and well-known fungus, the *ergot*. Italian rye-grass is the most liable to the ravages of this pest, and there are on record several cases in which ergotted rye-grass proved fatal to the animal fed upon it. Clover and the various leguminous plants appear more liable to the ergot disease than the natural grasses (except rye-grass), but I have on several occasions noticed this fungus on the spikelets of *Hordeum pratense, Festuca pratense,* and *Bromus erectus*. It has also been noticed that rye-grass rapidly developed under the influence of liquid manure is so rank that young animals fed upon it are poisonously affected.

Alderman Mechi states that in July, 1864, ten out of his thirty Shorthorn calves died in consequence of eating the heads of Italian rye-grass, and that the survivors' health was seriously injured. He was also unfortunate with his lambs, which, during the same month, were folded on Italian rye-grass. "Four days ago," writes the Alderman, "it was sewaged, having been prior to the former growth also guanoed. In four days it had grown from four to five inches, was of an intense green, and pronounced to be, by sharp practical men, just the food for lambs. Well, we put on our lambs, taking care to do so in the evening, after they had been well fed. My bailiff accompanied them, and, within five minutes, turning accidentally round, he saw two of the lambs with their heads in the air staggering (stomach staggers it is called) and frothing at the mouth. He immediately saw the mischief, removed the lambs, and on their way back to a bare fold some of them vomited the Italian rye-grass that they had just eaten, accompanied by frothy slime; others brought it up during the night. Some of them trembled, gaped, and showed all the same symptoms that my calves had done, such as rapid pulse, &c. Two or three of them are rather queer to-day. I hope that Professor Simmonds or some capable person will tell us how this is? If we mow this grass, bring it home, and cut it into chaff, all which tends to heat or dry it, it becomes wholesome food. The same remarks apply in degree to very succulent tares. If the Italian grass is brought home and given long and quite fresh to the calves, it will kill them. It does not appear to injure old ewes as it does lambs or shearlings. The dry weather has something to do with it. In wet weather the evil is much diminished, or disappears."

It is probable that the juice of this poisonous herbage was extremely rich in matters only semi-organised, and perhaps abounded in the crude substances from which the vegetable tissues are elaborated. Such rank grass as this was should not be used until it has attained to a tolerably developed state: in mature plants the juices contain more highly organised matters than are found in young vegetables.

The *Sorghuo, or Holcus Saccharatus.*—This plant, introduced to the notice of the British farmer but a few years ago, is only grown in these countries in small quantities. It is very rich in sugar, and cattle relish it greatly. Its composition, according to Dr. Voelcker, is as follows :—

Water	81·80
Albuminous matters	1·53
Insoluble ditto	0·66
Sugar	5·85
Wax and fatty matter	2·55
Mucilage, pectin, and digestible matters	2·59
Indigestible woody fibre	4·03
Mineral matter	0·99
	100·00

The plants referred to in the above analysis were cut in September. It is found that the composition of the plant is very different at different seasons.

Green Rye is employed as a forage crop, for which purpose it is well adapted. It is about equal in nutritive power to clover. According to Dr. Voelcker its composition is as follows :—

Water	75·423
Flesh-formers	2·705
Fatty matter	0·892
Gum, pectin, sugar, &c.	9·134
Woody-fibre	10·488
Mineral matter	1·358
	100·000

Buckwheat is occasionally cut in a green state and used as food for stock. Its composition, according to Einhof and Crome, is as follows :—

Water	82·5
Nitrogenous compounds	0·2
Extractive matters	2·6
Starch, &c.	4·7
Cellulose	10·0
	100·0

Rape is one of our most valuable plants for stock feeding. Two varieties are cultivated in these countries—the summer rape (*Brassica Campestris oleifera*) and winter rape (*Brassica rapus*). The great utility of rape arises from the circumstance of its being generally obtained as a *stolen* crop; for otherwise it is not quite equal to other plants that might be substituted for it—cabbages, &c. This plant is very rich in oily matters, and has been found well adapted both for the feeding of cattle and the fattening of sheep. Its composition, according to Voelcker, is shown in this table :—

COMPOSITION OF GREEN RAPE.

Water	87·050
Flesh-formers	3·133
Fatty matters	0·649
Other respiratory substances	4·000
Woody fibre	3·560
Mineral matter (ash)	1·608
	100·000

With respect to the value of rape for the feeding of stock in spring, Mr. Rham makes the following remarks :—

If the crop is very forward it may be slightly fed off, but in general it is best to let it remain untouched till spring. In the end of March and the beginning of April it will be a great help to the ewes and lambs. It will produce excellent food till it begins to be in flower, when it should immediately be ploughed up. The ground will be found greatly recruited by this crop, which has taken nothing from it, and has added much by the dung and urine of the sheep. Whatever be the succeeding crop, it cannot fail to be productive; and if the land is not clean, the farmer must have neglected the double opportunity of destroying weeds in the preceding summer, and in the early part of spring. If the rape is fed off in time, it may be succeeded by barley or oats, with clover or grass seeds, or potatoes, if the soil is not too wet. Thus no crop will be lost, and the rape will have been a clear addition to the produce of the land. Any crop which is taken off the land in a green state, especially if it be fed off with sheep, may be repeated without risk of failure, provided the land be properly tilled; but where cole or rape have produced seed, they cannot be profitably sown in less than five or six years after on the same land. The cultivation of rape or cole for spring food cannot be too strongly recommended to the farmers of heavy clay soils.

The Mustard Plant is occasionally used as food for sheep, for which purpose its composition shows it to be well adapted. Voelcker's analysis proves it to be very rich, relatively, in muscle-forming elements and in mineral matters; it might, therefore be with advantage combined with food relatively deficient in these principles.

COMPOSITION OF FRESH MUSTARD.

Water	86·30
Albuminous matters	2·87
Non-nitrogenous matters (gum, sugar, oil, &c.)	4.40
Woody fibre	4·39
Ash	2·04
	100·00

The Prickly Comfrey has been recommended as a good forage plant. It yields an abundant crop—or rather crops, for it may be cut several times in the year. The plant is a handsome one, and it might combine the useful with the ornamental if it were cultivated on demesne or villa farms. Dr. Voelcker states its composition to be as follows:—

Water	88·400
Flesh-forming substances	2·712
Heat and fat-producing matters ...	6·898
Ash	1·990
	100·000

Chicory is used as a forage crop on the Continent, and Professor John Wilson surmises that it may yet be generally cultivated for this purpose in Great Britain. At present it is rarely grown except for the sake of its roots, which are used as partial substitutes for, or adulterants of, coffee.

COMPOSITION OF CHICORY, ACCORDING TO ANDERSON.

	Fresh roots.	Fresh leaves.
Water	80·58	90·94
Nitrogenous matters	1·72	1·01
Non-nitrogenous substances ...	16·39	6·63
Ash	1·31	1·42
	100·00	100·00

Yarrow (*Achillæa millefolium*) is usually regarded as a weed, but sheep are very fond of it, and when they can get it, never fail to eat it greedily. It possesses astringent properties. Some writers have recommended it as a good crop for warrens and sands. Its composition, according to Way, is as follows :—

DRIED YARROW.

Albuminous matter	10·34
Fatty matters	2·51
Starch, gum, &c.	45·46
Woody fibre	32·69
Mineral matter	9·00
	100·00

Melons and *Marrows* have been used, but to a very limited extent, as food for stock. Mr. Blundell advocates their use in seasons of drought. He states that he has obtained more than forty tons per acre of both melons and marrows. They are relished by horses, oxen, sheep, and pigs. Mr. Blundell's advocacy has not been attended with much success, but it would be desirable to give these vegetables a further trial.

Dr. Voelcker's analysis of the cattle melon shows that it contains :—

Water	92·98
Albuminous matters	1·53
Oil	·73
Sugar, gum, &c.	2·51
Fibre	1·65
Ash	·60
	100·00

The Cabbage.—The composition of the Drumhead Cabbage has been studied by Dr. Anderson. He found a larger proportion of nutriment in the outer leaves than in the "heart," and ascertained that the young plants were richer in nutriment than those more advanced in age. His results show the desirability of cultivating the open-leaved, rather than the compact varieties of this plant.

ANALYSIS OF THE CABBAGE.—BY DR. ANDERSON.

	Outer leaves.	Heart leaves.
Water	91·08	94·48
Compounds containing nitrogen	1·63	0·94
Compounds destitute of nitrogen, such as gum, sugar, fibre, &c....	5·06	4·08
Ash (mineral matter)	2·23	0·50
	100·00	100·00

According to Fromberg, the composition of the whole plant is as follows:—

Water	93·40
Nitrogenous, or flesh-forming compounds	1·75
Non-nitrogenous substances, such as gum, sugar, &c.	4·05
Mineral matter	0·80
	100·00

Dr. Voelcker, who has more recently analysed the cattle cabbage, furnishes us with the following details of its composition:—

COMPOSITION OF CABBAGE LEAVES (OUTSIDE GREEN LEAVES).

Water	83·72
Dry matter	16·28
	100·00

The fresh and the dry matter consisted of:—

	Fresh Matter.	Dry matter. Per cent.
*Protein compounds	1·65	10·19
Non-nitrogenous matter	13·38	82·10
Mineral matter	1·25	7·71
	16·28	100·00
* Containing nitrogen	·26	1·63

In the following table the results of a more elaborate analysis of the *heart* and inner leaves are shown:—

GREEN FOOD.

COMPOSITION OF HEART AND INNER LEAVES.

	In natural state.	Dry.
Water	89·42	
Oil	·08	·75
*Soluble protein compounds	1·19	11·24
Sugar, digestible fibres, &c.	7·01	66·25
Soluble mineral matter	·73	6·89
†Insoluble protein compounds	·31	2·93
Woody fibre	1·14	10·77
Insoluble mineral matter	·12	1·17
	100·00	100·00
* Containing nitrogen	·19	1·79
† Containing nitrogen	·05	·47

If I were asked what plant I considered the most valuable for forage, I certainly should pronounce an opinion in favor of cabbage. This crop yields a much greater return than that afforded by the Swedish turnip, and it is richer in nutritive matter. Cabbages are greedily eaten by sheep and cattle, and the butter of cows fed upon them is quite free from the disagreeable flavor which it so often possesses when the food of the animal is chiefly composed of turnips. If the cabbage admitted of storing, no more valuable crop could be cultivated as food for stock.

Mr. John M'Laren, of Inchture, Scotland, gives in the "Transactions of the Highland Agricultural Society of Scotland for 1857," a report on the feeding value of cabbage, which is highly favorable to that plant :—

On the 1st December, 1855 (says the reporter), two lots of Leicester wethers, bred on the farm, and previously fed alike, each lot containing ten sheep, were selected for the trial by competent judges, and weighed. Both lots were put into a field of well-sheltered old lea, having a division between them. All the food was cut and given them in troughs, three times a day. They had also a constant supply of hay in racks.

At the end of the trial, on the 1st of March, 1856, the sheep were all re-weighed, sent to the Edinburgh market, and sold same day, but in their separate lots. As I had no opportunity of getting the dead weights, I requested Mr. Swan, the salesman, to give his opinion on their respective qualities. This was to the effect that no difference existed in their market value, but that the sheep fed on turnips would turn out the best quality of mutton, with most profit for the butcher. Both lots were sold at the same price, viz., 52s. 6d. During the three months of trial, we found that each lot consumed about the same weight of food—viz., 8 tons 13 cwt. 47 lb. of cabbage, being at the rate of 21⅓ lbs. per day for each sheep, and 8 tons 10 cwt. 7 lb. Swedes, being at the rate of 20$\frac{8}{10}$ lb. per day.

It will be seen, by referring to the table (see next page), that in this trial the Swede has proved of higher value for feeding purposes than the cabbage, making 11 st. 4 lb. of gain in weight, whilst the cabbage made 10 st. 9 lb. At the same time, 3 cwt. 40 lb. less food were consumed; and taking the mutton gained at 6d. per lb., the Swedes consumed become worth 9s. 3¼d. per ton, while the gain on the cabbage, at the same rate, makes them worth 8s. 7d. per ton. But from the great additional weight of the one crop grown over the other, the balance, at the prices, &c., mentioned, is in favor of the cabbage by £1 15s. 11¾d. per acre.

These results certainly speak strongly in favor of the cabbage; but the weight of the acreable crop of cabbages stated in the table appears to be unusually great. So heavy a crop is rarely obtained.

Furze (*Gorse, or Whins*).—Notwithstanding the natural historical knowledge of Goldsmith, his poetical description of the furze is far from accurate. This plant, instead of being "unprofitably gay," deserves to rank amongst the most valuable vegetables cultivated for the use of the domestic animals. It grows and flourishes under conditions which most injuriously affect almost every other kind of fodder and green crop. Prolonged drought in spring and early summer not unfrequently renders the hay crop a scanty one; while autumn and winter frosts change the nutriment of the mangels and turnips into decaying and unwholesome matter. Under such circumstances as these, the maintenance of cattle in good condition is very expensive, unless in places where a supply of furze is available. This plant is rather improved than otherwise by exposure to a temperature which would speedily destroy a mangel or a turnip; and, although it thrives best when abundantly supplied with rain, it can survive an exceedingly prolonged drought without sustaining much injury.

GREEN FOOD.

TABLE

Showing the Difference of Weight grown on an Acre of Cabbage and an Acre of Swedes, and the Value of each for Feeding.

No. of Sheep in each lot.	Kinds of Food.	Weight of Ten Sheep, 1st Dec., 1855.	Weight of Ten Sheep, 1st Mar., 1856.	Gain.	Value of Gain, taking Mutton at 6d. per lb.	Total Weight of Food consumed in Three Months by each lot.	Value of Food consumed per Ton.	Total Weight per Acre of each Crop.	Value of each Crop per Acre.	Extra Cost on each Crop per Acre.	Free Value of each Crop per Acre.	Balance in favor of Cabbage per Acre.
		st. lb.	st. lb.	st. lb.	£ s. d.	tons. cwt. lb.	s. d.	tons. cwt.	£ s. d.	£ s. d.	£ s. d.	£ s. d.
10	Cabbage	90 10	101 5	10 9	3 14 6	8 13 47	8 7	42 14	18 6 6	4 10 11	13 15 7	1 15 11¾
10	Swedes	89 3	100 7	11 4	3 19 0	8 10 7	9 3¼	26 12	12 6 7¼	0 7 0	11 19 7¼	

The furze is a member of the family *Leguminosæ*, which includes so many useful plants, such as, for example, the pea, the bean, and the clovers. There are three varieties of it met with in this country—namely, the common furze, *Ulex europæus*, the dwarf furze, *Ulex nanus*, and the Irish, or upright furze, *Ulex strictus*.

The common furze is a hardy shrub, and grows luxuriantly at an elevation far higher than the limits of cereal cultivation. It flourishes on any kind of soil which is moderately dry, and heavy crops may easily be raised on uplands almost incapable of producing grass. The dwarf furze is never cultivated, but as it grows at a still greater elevation, and on a poorer soil than the larger varieties, it might be profitably cultivated on very high uplands. The Irish furze yields a softer and less prickly food than the other kinds, but as it does not usually bear seed, and must therefore be propagated by cuttings, its cultivation has hitherto been limited to but a few localities.

The produce of an acre of furze appears to be at least equal to that of an acre of good meadow. The Rev. Mr. Townsend of Aghada, county of Cork—the most zealous and successful advocate for the cultivation of this plant—informed me that he had obtained so much as 14 tons per acre; a fact which proves that the furze is a plant which is well deserving of the attention of the farmer.

Furze is an excellent food for every kind of stock. Cattle, although they may at first appear not to relish its prickly shoots, soon acquire a fondness for it. I have known several instances of herds being fed almost if not entirely on the bruised plant, and to keep in good condition. The late Professor Murphy, of Cork, stated that on the farm of Mr. Boulger, near Mallow, thirty-five cows were fed on crushed furze, which they "devoured voraciously." Each animal received daily from four to six stones of the crushed plant, to which were added a little turnip pulp and a small quantity of oats. The milk and butter yielded by these cows were considered excellent. In a letter addressed to me by a very intelligent feeder, Mr. John Walsh,* of Stedalt, county of Dublin, the following remarks in relation to this subject are made :—

I had lately an opportunity of seeing a herd of cattle of about sixty head, of which twenty had been fed with furze prepared with my machine for about six weeks before being put out to grass. The condition of these was so superior that I pointed out every one of them, one after the other, out of the herd. The owner of the cattle had made the same observation; it was new to him but not to me.

Furze is seldom given to sheep or pigs, but I believe that it might with advantage enter into the dietary of those animals. Some of my friends who have lately tried it with pigs report favorably as to its effects. Horses partly fed upon this plant keep in good condition; it is usually given to them cut merely into lengths of half an inch or an inch, but it would be better to give it to them finely bruised. A horse during the night will eat a much larger quantity of coarsely cut furze than of the well bruised article, because he is obliged to expend a great deal of muscular power in bruising the furze, and must, consequently, use an additional quantity of the food to make up for the corresponding waste of tissue.

Until quite recently, the chemistry of the furze was very little studied. The analysis of this plant made many years ago by Sprengel gave results which, in the present advanced condition of agricultural chemistry, are quite valueless. The late Professor Johnston merely determined its amount of water, organic matter, and ash. I believe I was the first to make a complete investigation into the composition of this plant according to the methods of modern chemical analysis. I made two examinations. The first was of shoots cut on the 25th April, 1860, on the lands of Mr. Walsh of Stedalt, near Balbriggan, in the county of Dublin. The shoots were, in great part, composed of that year's growth, with a small proportion of the

---

* This gentleman has invented an exceedingly simple but effective furze-bruiser, which I hope soon to see in general use.

shoots of the previous year. They were very moist, and their spines, or thorns, were rather soft. Their centesimal composition was as follows:—

Water	78·05
Nitrogenous, or flesh-forming principles	2·18
Fat-forming principles (oil, starch, sugar, gum, &c.)	8·20
Woody fibre	10·17
Mineral matter (ash)	1·40
	100·00

The second analysis was made of furze cut on the 15th August, 1862. The following were the results obtained:—

Water	72·00
Nitrogenous, or flesh-forming principles	3·21
Oil	1·18
Other fat-forming principles (starch, gum, &c.) ...	8·20
Woody fibre	13·33
Mineral matter	2·08
	100·00

The specimen was allowed to lie for a few days in a dry room, so that it lost a little water whilst in my possession, before it was subjected to analysis.

The sample cut in August contained a larger amount of nutriment than the specimen analysed in the spring; but its constituents appeared to be much less soluble in water, and therefore, less digestible.

Professor Blyth, of the Queen's College, Cork, has more recently made a very elaborate analysis of furze, grown in the county of Cork, which gave results still more favorable to the plant than those arrived at by me—probably because the specimens furnished to him were drier than mine.

GREEN FOOD.

ANALYSIS OF FRESH FURZE, BY DR. BLYTH.

100 parts contain :—
Matters readily soluble in water and easily digested.

*Albuminous, or flesh-forming compounds	1·68
Fat and heat-producing, or respiratory elements, viz., sugar, gum, &c. &c.	7·83
Ash	0·83
Total matters soluble in water	10·34
*Containing nitrogen	0·265

Matters insoluble in water.

Oil	2·14
*Albuminous, or flesh-producing compounds	2·83
Fat and heat-producing, or respiratory elements	1·00
Woody fibre	28·80
Ash	3·23
Total matters insoluble in water	38·00
Water, expelled at 212...	51·50
	99·48
Total nitrogen in plant	0·71
Total albuminous, or flesh-producing compounds	4·51
Total respiratory, or heat and fat-producing compounds	8·83
Total ash...	4·06

The ash contains in 100 parts :—

Potash	20·00
Phosphoric acid	8·72
*Containing nitrogen	0·445

If the large per-centage of water be deducted, the dry, nutritive matters can then be more readily compared with the amount of the same substances in other feeding articles :—

Composition of 100 parts of furze dried at 212°. Matters soluble in water in the dry furze.

*Albuminous compounds	3·47
Respiratory elements	16·15
Ash	1·71
Total matters soluble in water...	21.33
*Containing nitrogen	0·546

Matters insoluble in water in the dry furze.

Oil	4·41
*Albuminous compounds	5·84
Respiratory elements	2·06
Woody fibre	59·38
Ash	6·66
Total matters insoluble in water	78·35
	99·68
Total nitrogen in dry furze	1·46
Total albuminous compounds	9·13
Total respiratory elements	18·20
Total ash	8·36
*Containing nitrogen	0·917

Composition of ash per cent.

Potash	20·00
Phosphoric Acid	8·72

The results of these analyses show that dry furze contains an amount of nutriment equal to that found in dry grass. The nature of its composition resembles, as might be expected, that of its allied plants, vetches, &c., and therefore it exceeds the grasses in its amount of ready formed fatty matter.

SECTION IV.

STRAW AND HAY.

Straw.—At the present time, when the attention of the farmer is becoming more and more devoted to the production of meat, it is very desirable that his knowledge of the exact nutritive value of the various feeding substances should be more extensive than it is. No doubt, most feeders are practically acquainted with the relative value of corn and oil-cake—of Swedish turnips and white turnips; but their knowledge of the food equivalents of many other substances is still very defective. For example, every farmer is not aware that Indian corn is a more economical food than beans for fattening cattle, and less so for beasts of burthen. Locust-beans, oat-dust, malt-combings, and many other articles, occasionally consumed by stock, have not, as yet, determinate places assigned to them in the feeder's scale of food equivalents.

The points involved in the economic feeding of stock are not quite so simple as some farmers, more especially those of the amateur class, appear to believe. There are many feeders who sell their half-finished cattle at a profit, and yet they cannot, without loss, convert their stock into those obese monsters which are so much admired at agricultural shows. The complete fattening of cattle is a losing business with some feeders, and a profitable one with others. Stall-feeding is a branch of rural economy which, perhaps more than any other, requires the combination of "science with practice;" yet how few feeders are there who have the slightest knowledge of the composition of food substances, or who are agreed as to the feeding value, absolute or relative, of even such well-known materials as oil-cake, straw, or oats! "It is thus seen how inexact are the equivalents which are understood to be established for the different foods used for the maintenance of the animals. It is equally plain, when we reflect on the different methods pursued for the preservation of the animals, that we are still far from having attained that perfection towards which our efforts tend. Visit one hundred farms, taken by chance in different parts of the country, and you will find in each, methods directly opposite —a totally peculiar manner of managing the stalls; you will see, in short, that the conditions of food, of treatment, and of hygiene, remain not understood in seven-eighths of rural farms."*

The straws of the cereal and leguminous plants are a striking illustration of the erroneous opinions and practices which prevail amongst agriculturists with respect to particular branches of their calling. The German farmers regard straw as the most valuable constituent of home-made fertilisers, and

* H. Le Docte, in *Journal de la Société Centrale d'Agriculture de Belgique*.

their leases in general prohibit their selling off the straw produced on their farms. Yet chemical analysis has clearly proved that the manurial value of straw is perfectly insignificant, and that, as a constituent of stable manure, it is chiefly useful as an absorbent of the liquid egesta of the animals littered upon it. As food for stock, straw was at one time regarded by our farmers as almost perfectly innutritious; some even went so far as to declare that it possessed no nutriment whatever, and even those who used it, did so more with the view of correcting the too watery nature of turnips, than with the expectation of its being assimilated to the animal body. Within the last few years, however, straw has been largely employed by several of the most intelligent and successful feeders in England, who report so favorably upon it as an economical feeding stuff, that it has risen considerably in the estimation of a large section of the agricultural public. Now, even without adopting the very high opinion which Mechi and Horsfall entertain relative to the nutritive power of straw, I am altogether disposed to disagree with those who affirm that its application should be restricted to manurial purposes. Unless under circumstances where there is an urgent demand for straw as litter, that article should be used as food for stock, for which purpose it will be found, if of good quality, and given in a proper state, a most economical kind of dry fodder—equal, if not superior to hay, when the prices of both articles are considered.

The composition of straw is very different from that of grain. The former contains no starch, but it includes an exceedingly high proportion of woody fibre; the latter is in great part composed of starch, and contains but an insignificant amount of woody fibre. Dr. Voelcker, the consulting chemist to the Royal Agricultural Society of England, and Dr. Anderson, chemist to the Highland and Agricultural Society of Scotland, have made a large number of analyses of the straws of the cereal and leguminous plants, the results of which are of the highest interest to the agriculturist. In the following tables the more important results of these investigations are given:—

ANALYSES OF STRAW, BY DR. VOELCKER.

	No. 1. Wheat, just ripe and well harvested.	No. 2. Wheat, over ripe.	No. 3. Barley, dead ripe.	No. 4. Barley, not too ripe.	No. 5. Oat, cut green.	No. 6. Oat, cut when fairly ripe.	No. 7. Oat, over ripe.	No. 8. Bean.	No. 9. Pea.	No. 10. Flax Chaff.
Water	13·33	9·17	15·20	17·50	16·00	16·00	16·00	19·40	16·02	14·60
Albumen, and other protein compounds:—										
a. Soluble in water	1·28	0·06	0·68	} 5·73	5·51	2·62	1·29	1·51	3·96	} 4·75
b. Insoluble in water	1·65	2·06	3·75		2·98	1·46	2·36	1·85	5·90	
	1·74	0·65	1·36	1·17	1·57	1·05	1·25	1·02	2·34	2·82
Oil	4·26	3·46	2·24	} 71·44	16·04	10·57	3·19	4·18	8·32	8·72
Sugar, mucilage, extractive matters, &c. (soluble in water)	19·40	} 82·26	5·97		26·34	30·17	27·75	2·75	17·74	18·56
Digestible woody fibre and cellulose										
Indigestible fibre, &c.	54·13		66·54		24·86	31·78	41·82	65·58	42·79	43·12
Inorganic matter:—										
a. Soluble ...	1·13	1·29	2·88	} 4·52	5·76	3·64	2·26	2·31	2·72	4·07
b. Insoluble ...	3·08	1·05	0·38		0·94	2·71	4·08	1·40	2·21	3·36
	100·00	100·00	100·00	100·00	100·00	100·00	100·00	100·00	100·00	100·00

*** This table contains in a condensed form all the results of Voelcker's analyses of the straws which are given in his paper published in the *Journal of the Royal Agricultural Society of England*, vol. xxii. part 2. 1862.

Nos. 5, 6, and 7 were analysed shortly after being cut, when they contained a high proportion of water. They have, therefore, been calculated to contain 16 per cent. of moisture so as to arrive at accurate relative results.

ANALYSES OF STRAW, BY DR. ANDERSON.

	Wheat from East Lothian.	Wheat from Kent.	Barley from East Lothian.	Barley from Kent.	Sandy Oat from East Lothian.		Oat from Sea level, East Lothian.	Oat from 850 feet above Sea level, East Lothian.	Oat from Mellhill, Inchture, Scotland.	Oat from Kent (White one side).
Water	10·62	11·15	11·44	11·15	11·10	11·70	10·95	12·60	11·28	11·70	10·55
Flesh-formers—											
Soluble	0·86	0·37	1·37	1·42	0·39	0·66	0·40	1·03	0·67	0·92	0·95	0·33
Insoluble	0·51	1·12	1·00	1·54	1·12	1·98	0·93	0·43	0·38	0·39	1·21	0·33
Oil	0·80	1·00	1·50	0·97	0·88	1·05	1·45	0·77	1·25	1·36	1·60	1·00
Respiratory elements—											
Soluble	2·68	6·68	5·26	3·22	6·11	4·56	10·12	6·90	7·16	7·42	12·01	6·23
Insoluble	44·83	36·43	38·79	35·56	38·38	27·95	33·52	34·77	24·28	29·55	23·35	30·95
Woody fibre	32·88	34·78	35·01	41·34	36·62	47·53	35·36	38·73	48·49	44·40	45·27	47·40
Ash	6·20	8·04	6·32	4·21	5·62	4·85	6·36	6·28	5·11	5·07	3·95	3·62
	99·43	99·35	100·40	99·70	100·27	99·68	99·84	99·86	99·94	100·39	100·14	100·41

** This table is compiled from Dr. Anderson's paper in the *Transactions of the Highland and Agricultural Society of Scotland* for March, 1862.

Many very important conclusions are deducible from the facts recorded in these valuable tables. We learn from them that straw is more nutritious when it is cut in the ripe state than when it is permitted to over-ripen, and that *green* straw contains a far greater amount of nutriment than is found even in the ripe article. It appears also that the least nutritious kind of straw equals the best variety of turnips in its amount of flesh-forming principles, and greatly exceeds them in its proportion of fat-forming elements. We further learn that in general the different kinds of straw will be found to stand in the following order, the most nutritious occupying the highest, and the least nutritious the lowest place :—

1. Pea-haulm.
2. Oat-straw.
3. Bean-straw with the pods.
4. Barley-straw.
5. Wheat-straw.
6. Bean-stalks without the pods.

It is a matter to be regretted that we possess so little accurate knowledge of the chemical composition of the plants cultivated in Ireland. No doubt the analyses of English grown wheat, beans, mangels, and other plants, serve to give us a general idea of the nature of those vegetables when produced in this country. But this kind of information, though very important, must necessarily be defective, as differences in climate modify—often to a considerable extent—the composition of almost every vegetable. Thus, the results of Anderson's analyses prove Scotch oats to be superior, as a feeding stuff, to Scotch barley, whilst, according to Voelcker and the experience of most English feeders, the barley of parts of England is superior to its oats. It follows, then, that whilst the results of the analyses of straw, made by Voelcker and Anderson are of great interest to the Irish farmer, they would be still more important to him had the straw to which they relate been the produce of Irish soil. In order, therefore, to enable the Irish farmer to form a correct estimate of the value of his straw, we should put him in possession of a more perfect knowledge of its composition than that which is derivable from the investigations to which I have referred. The straws of the cereals—which alone are used here to any extent—should be analysed as carefully and as frequently as those of Great Britain have been; and if such were done, I have no doubt but that the results would indicate a decided difference in composition between the produce of the two countries. Some time ago I entered upon what, at the time, I had intended should be a complete investigation into the composition of Irish straws; but which want of time prevented me from making more than a partial one. The results are given in the following tables:—

ANALYSES OF IRISH OAT-STRAW.

	No. 1. From Co. Wicklow.	Obtained in the Dublin Market.		
		No. 2.	No. 3.	No. 4.
Water	14·00	14·00	14·00	14·00
Flesh-forming principles—				
a. Soluble in water	4·08	2·02	2·04	1·46
b. Insoluble in water	2·09	3·16	3·00	2·23
Oil	1·84	1·40	1·26	1·00
Sugar, gum, and other fat-forming matters	13·79	12·67	10·18	11·16
Woody fibre	59·96	61·79	65·45	65·29
Mineral matter	4·24	4·96	4·07	4·86
	100·00	100·00	100·00	100·00

All the specimens of oats, the analyses of which are given in the preceding table, are assumed to contain 14 per cent. of water, in order the more correctly to compare their nutritive value. No. 1 contained 18·23 per cent. of water; No. 2, 12·90; No. 3, 12·74; and No. 4, 12·08. Oat straw, before its removal from the field, often contains nearly half its weight of water; but after being for some time stacked, the proportion of moisture rarely exceeds 14 per cent.

ANALYSES OF IRISH WHEAT-STRAW.

	No. 1. Green, changing to yellow. County Kildare.	No. 2. Ripe. County Dublin.	No. 3. Over Ripe. County Dublin.	Obtained in the Dublin Markets.		
				No. 4.	No. 5.	No. 6.
Water	13·00	13·15	12·14	10·88	11·22	12·12
Flesh-forming principles—						
a. Soluble in water	1·25	0·98	0·44	0·06	0·42	0·30
b. Insoluble in water	1·26	1·40	1·41	1·90	1·00	1·76
Oil	1·22	1·13	1·14	0·90	1·17	1·08
Sugar, gum, and other fat-forming matters	4·18	3·98	3·88	4·08	3·89	4·30
Woody fibre	75·84	76·17	77·76	78·67	79·18	77·15
Mineral matter (ash).	3·25	3·19	3·23	3·51	3·12	3·29
	100·00	100·00	100·00	100·00	100·00	100·00

The results of these analyses are somewhat different from those arrived at by Voelcker and Anderson. They show that properly harvested Irish oat and wheat straws are far more valuable than those of Scotland, and somewhat less nutritive than those produced in England. They also show that wheat-straw is allowed to over-ripen, by which a very large proportion of its nutritive principles is eliminated and altogether lost, and a considerable part of the remainder converted into an insoluble, and therefore less easily digestible state. Nor is there any advantage to the grain gained by allowing it to remain uncut after the upper portion of the stem has changed from a green to a yellowish color; on the contrary, it also loses a portion—often a very considerable one—of its nitrogenous, or flesh-forming constituents. It has been clearly proved that wheat cut when green, yields a greater amount of grain, and of a better quality too, than when it is allowed to ripen fully; yet, how often do we not see fields of wheat in this country allowed to remain unreaped for many days, and even weeks, after the crop has attained to its full development!

The oat-straw obtained in the Dublin Market proved less

valuable than the green straw which I selected myself from a field of oats; but the discrepancy between them was far less than between the nearly ripe wheat-straw and the straw of that plant purchased in Dublin. During visits which I have paid in harvest-time to the North of Ireland, I noticed that the oats were generally cut whilst green, whereas wheat was almost invariably left standing for at least a week after its perfect maturation, probably for the following reasons:—Firstly, because oats are more liable to shed their seed; secondly, because there is a greater breadth of that crop to be reaped, which necessitates an early beginning; and, lastly, because most farmers know that over-ripe oat-straw is worth but little for feeding purposes, as compared with the greenish-yellow article.

As compared with white turnips, the nutritive value of oat-straw stands very high, for whilst the former contains but little more than 1 per cent. of flesh-formers, and less than 5 per cent. of fat-formers, the latter includes about 4 per cent. of flesh-formers, and 13 per cent. of fat-formers. Again, whilst the amount of woody fibre in turnips is only about 3 per cent., that substance constitutes no less than 60 per cent. of oat-straw. In comparison with hay—taking into consideration the prices of both articles—oat-straw also stands high, as will be seen by comparing the following analyses of common meadow hay with that of properly harvested straw:—

	Meadow Hay.	Oat Straw.
Water	14·61	14·00
Flesh-forming constituents	8·44	6·17
Respiratory and fatty matters	43·63	15·63
Woody fibre	27·16	59·96
Mineral matter (ash)	6·16	4·24
	100·00	100·00

Woody fibre is as abundant a constituent of the straw of the cereals as starch is of their seeds, and if the two substances were equally digestible, straw would be a very valuable food—

superior even to the potato. At one time it was the general belief that woody fibre was incapable of contributing in the slightest degree to the nutrition of animals, but the results of recent investigations prove that it is, to a certain extent, digestible. In the summer of 1859 two German chemists, Stöckhardt and Sussdorf, made a series of experiments, with the view of ascertaining whether or not the cellulose* of the food of the sheep is assimilated by that animal. The results of this inquiry are of importance, seeing that they clearly prove that even the hardest kind of cellulose—*sclerogen*, in fact—is capable of being assimilated by the Ruminants. The animals selected were two wethers, aged respectively five and six years. They were fed—firstly, upon hay alone; secondly, upon hay and rye-straw; thirdly upon hay and the sawdust of poplar wood, which had been exhausted with lye (to induce the sheep to eat the sawdust, it was found necessary to mix through it some rye-bran and a little salt); fourthly, hay and pine-wood sawdust, to which was added bran and salt; fifthly, spruce sawdust, bran and salt; sixthly, hay, pulp of linen rags (from the papermaker), and bran. The experiments were carried on from July till November, excepting a short time, during which the animals were turned out on pasture-land, to recover from the injurious effects of the fifth series of experiments—produced probably by the resin of the spruce. The animals, together with their food, drink, and egesta, were weighed daily. The amount of cellulose in the food was determined, and the proportion of that substance in the egesta was also ascertained; and as there was a considerable discrepancy between the two amounts, it was evident that the difference represented the

* Cellulose is the term applied to the chemical substance which forms woody fibre. The latter is made up of very minute spindle-shaped tubes, In young and succulent plants these tubes are often lined with layers of soft cellulose. In many plants—such as trees—in a certain stage of development, the substance lining the cells is very hard, and is termed *lignin*, or *sclerogen*. This substance is merely a modification of cellulose; and both resemble in composition sugar and starch so closely that, by heating them with sulphuric acid, they may be converted into sugar.

weight of the cellulose assimilated by the animals. In this way it was ascertained that from 60 to 70 per cent. of the cellulose of hay, 40 to 60 per cent. of the cellulose of straw, 45 to 50 per cent. of the cellulose of the poplar wood, 30 to 40 per cent. of the cellulose of the pine, and 80 per cent. of the cellulose of the paper pulp was digested.

In stating the results of his analyses of the straws, Professor Voelcker sets down as "digestible" that portion of the cellulose which he found to be soluble in dilute acids and alkaline solutions; but he admits that the solvents in the stomach might dissolve a larger amount. The results of the experiments of Stöckhardt and Sussdorf prove that 80 per cent. of the cellulose of paper (the altered fibre of flax) is assimilable, and it is, therefore, not unreasonable to infer that the cellulose of a more palatable substance than paper might be altogether digestible.

The facts which I have adduced clearly prove that the straws of the cereals possess a far higher nutritive power than is commonly ascribed to them; that when properly harvested they contain from 20 to 40 per cent. of undoubted nutriment; and lastly, that it is highly probable that their so-called indigestible woody fibre is to a great extent assimilable.

The composition of cellulose is nearly, if not quite, identical with that of starch, and it may therefore be assumed to be equal in nutritive power to that substance—that is, it will, if assimilated, be converted into four-tenths of its weight of fat. Now as cellulose forms from six-tenths to eight-tenths of the weight of straws, it is evident that if the whole of this substance were digestible, straws would be an exceedingly valuable fattening food. When straw in an unprepared state is consumed, there is no doubt but that a large proportion of its cellulose remains unappropriated—nay more, it is equally certain that the hard woody fibre protects, by enveloping them, the soluble and easily digestible constituents of the straw from the action of the *gastric juice*. I would, therefore, recommend that straw should be either cooked or fermented before being made use of; in either of these states its constituents are far more diges-

tible than when the straw is merely cut, or even when it is in the form of chaff. An excellent mode of treating straw is to reduce it to chaff, subject it to the action of steam, and mix it with roots and oil-cake or corn. Mr. Lawrence, of Cirencester, one of the most intelligent agriculturists in England, cooks his chaff, which he largely employs, in the following manner:—
"We find that, taking a score of bullocks together fattening, they consume, per head per diem, 3 bushels of chaff mixed with just half a hundred-weight of pulped roots, exclusive of cake or corn; that is to say, rather more than 2 bushels of chaff are mixed with the roots, and given at two feeds, morning and evening, and the remainder is given with the cake, &c., at the middle day feed, thus:—We use the steaming apparatus of Stanley, of Peterborough, consisting of a boiler in the centre, in which the steam is generated, and which is connected by a pipe on the left hand with a large galvanised iron receptacle for steaming food for pigs, and on the right with a large wooden tub lined with copper, in which the cake, mixed with water, is made into a thick soup. Adjoining this is a slate tank of sufficient size to contain one feed for the entire lot of bullocks feeding. Into this tank is laid chaff, about one foot deep, upon which a few ladles of soup are thrown in a *boiling state;* this is thoroughly mixed with the chaff with a three-grained fork, and pressed down firm; and this process is repeated until the slate tank is full, when it is covered down for an hour or two before feeding time. The soup is then found entirely absorbed by the chaff, which has become softened, and prepared for ready digestion." A cheap plan is to mix the straw with sliced roots, moisten the mass with water, and allow it to remain until a slight fermentation has set in. This process effectually softens and disintegrates, so to speak, the woody fibre, and sets free the stores of nutritious matters which it envelopes. Some farmers who hold straw in high estimation, prefer giving it just as it comes from the field; they base this practice on the belief that Ruminants require a bulky and solid food, and that their digestive powers are quite sufficient to effect

the solution of all the useful constituents of the straw. It may be quite true that cattle, as asserted, can extract more nutriment out of straw than horses can, but that merely proves the greater power of their digestive organs. No doubt the food of the Ruminants should be bulky; but I am quite sure that cooked or fermented straw is sufficiently so to satisfy the desire of those animals for quantity in their food.

So far as I can learn, all the carefully conducted feeding experiments to test the value of straw which have been made, have yielded results highly favorable to that article. Mr. Blundell, in a paper on "The Use and Abuse of Straw," read before the Botley (Hampshire) Farmer's Club, states that in his experience he found straw to be more economical than its equivalent of roots or oil-cake, in the feeding of all kinds of cattle:—

> I find (says Mr. Blundell) that dairy cows, in the winter months, if fed on large quantities of roots, particularly mangels and carrots, will refuse to eat straw almost entirely, and become very lean; but they will always eat a full portion of sweet, well-harvested straw, when they get a small and moderate allowance of roots, say, for an ordinary-sized cow, 15 lbs. of mangel three times per day, the roots being given whole, just in the state they come from the store heap. Again, calves and yearlings being fed with roots in the same way, will eat a large quantity of straw, and when they have been kept under cover I have had them in first-rate condition for many years past. Also, in fattening beasts, when they get a fair allowance of roots, say 65 to 70 lbs. per day, with from 3 to 4 lbs. of cake or meal in admixture, they will eat straw with great avidity, and do well upon it, and make a profit. It is, however, often the case that bullocks receive 100 lbs., or upwards, of roots per day, with a large quantity of cake or meal, often 10 or 12 lbs. per day; they will not then look at straw, and are obliged to be fed with hay. The cost price of these quantities and kinds of food stands so high that the animals do not yield a profit; for although they may make meat a little faster, yet the proportionate increase is nothing compared to the increased cost of the feeding materials used.

Mr. Blundell gives us also the tabulated results of one of his experiments, which prove that by the use of straw there is to be obtained something more than manure by the feeding of stock:—

COST OF FEEDING AN OX PER WEEK WITH STRAW, ETC.,
ACCORDING TO MR. BLUNDELL.

	s.	d.
4 lbs. of oil-cake per day, or 38 lbs. per week, at £10 per ton	2	6
64 lbs. of roots ditto, or 4 cwt. ditto, at 13s. 4d. ditto...	2	8
20 lbs. of straw feeding, or 1¼ cwt. ditto, at 30s. ditto...	1	10½
20 lbs. of straw litter, or 1¼ cwt. ditto, at 15s. ditto ...	0	1½
Attendance, &c., per week ...	0	1
	8	0½
Deduct value of manure, per week ...	1	3½
	6	9
Increased value of ox per week	10	0
Deduct cost of feeding	6	9
	3	3

If we now turn to the study of the composition of straw regarded from an economic point of view, we shall find that the theoretical deductions therefrom harmonise with the results of actual feeding experiments. Let us assume that 100 parts of oat-straw contain on an average—

 1 part of oil,
 4 parts of flesh-formers,
 10 parts of sugar, gum, and other fat-formers, and
 30 parts of digestible fibre;

and if the price of the straw be 30s. per ton, we shall have at that cost the following quantities of digestible substances:—

ONE TON OF OAT-STRAW, AT 30s., CONTAINS:—

	lbs.
*Oil ...	22·4
Flesh-forming principles ...	89·6
Sugar, gum, and other fat-forming substances ...	224·0
Digestible fibre ...	672·0
	1,008·0
†Total amount of fat-formers, calculated as starch ...	952·0
Add flesh-formers ...	89·6
Total amount of nutritive matter ...	1,041·6

* One part of oil is equal to 2½ parts of starch—that is, 2½ parts of starch are expended in the production of 1 part of fat.
† No difference is here assumed between the nutritive value of sugar and starch.

We shall now compare this table with a similar one in relation to the composition of linseed cake, which will place the greater comparative value of straw in a clearer light.

A fair sample of linseed-cake contains, centesimally—

Flesh-formers	26
Oil	12
Gum, mucilage, sugar, &c.	34
Woody fibre	6

ONE TON OF LINSEED CAKE, AT £11, CONTAINS:—

	lbs.
Flesh-forming principles	582·4
Oil	268·8
Gum, sugar, and other fat-formers	761·6
Woody fibre	74·4
	1,687·2
Total amount of fat-formers, calculated as starch	1,508·0
Add flesh-formers	582·4
Total amount of nutriment	2,090·4

These comparisons are very instructive and important. We learn from them that we pay £11 for 2,000 lbs. of nutriment, when we purchase a ton of linseed-cake, whereas, when we invest 30s. in a ton of straw, we receive 1,000 lbs. of digestible aliment. It cannot be said that I have strained any points in favour of the straw; on the contrary, I believe that when that article is cut in proper season and well harvested, its composition will be found far superior to that detailed in the comparative analysis. It must be borne in mind, too, that I take no account of the 30 per cent. of the so-called indigestible woody fibre which straw contains, and which, I believe, is partly assimilable under ordinary circumstances, and could be rendered nearly altogether digestible by proper treatment; on the other hand, I have assumed that the woody fibre of the oil-cake is completely digestible, although I believe it is in reality less so than the fibre of straw.

It is an important point in the composition of oil-cakes, that they contain a large proportion of ready-formed fatty matters which can, with but little alteration, be at once transmuted into animal fat. There are some individuals of the genus *Homo* to whose stomachs fat, *per se*, is intolerable; nevertheless, as a general rule, fatty substances exercise a favorable influence in the process of digestion, and, either in a separate state, or intimately commingled with other aliments, constitute a large proportion of the food of man. Digestion in the lower animals is, no doubt, similarly promoted by mixing with the aliments which are to be subjected to that process, a due proportion of oily or fatty matter. Straw is relatively deficient in the flesh-forming principles, and abounds in the fat-forming elements—of which, however, the most valuable, oil, is the least abundant. Now, if we add to straw a due proportion of some substance very rich in flesh-formers and oil, the compound will possess in nicely adjusted proportions all the elements of nutrition. Perhaps the best kind of food which we could employ for this purpose is linseed meal. It contains about 24 per cent. of flesh-formers, 35 per cent. of a very bland oil, and 24 per cent. of gum, sugar, and mucilage. Linseed-cake may be substituted for linseed-meal; but the meal, though its cost is 15 per cent. greater, is, I believe, rather the better article of the two. Its flesh-formers are more soluble, and its oil thrice more abundant and far more palatable than the same principles in most samples of oil-cake. An important point, too, is, that linseed, unlike linseed-cake, is not liable to adulteration. As linseed possesses laxative properties it cannot be largely employed; the addition, however, of bean-meal—the binding tendency of which is well known—to a diet partly composed of linseed will neutralise, so to speak, the relaxing influence of the oily seed. If oil-cakes be used as an adjunct to straw, rape-cake will be found more economical than linseed-cake. If it be free from mustard, well steamed, and flavored with a little treacle, or a small quantity of locust-beans, it will be readily consumed, and even relished, by dairy and fattening stock.

Hay.—There is no food substance more variable or more complex than hay, for under that term are included, not only mixtures of grasses, but also of leguminous plants—clover, for example. The herbage of no two meadows is exactly alike; and the composition of the meadow plants is so greatly modified by differences of climate, soil, and mode of culture, that we have nothing to excite our wonder in the extreme variability of hay.

The composition of the hay made from clover, lucerne, and various other kinds of artificial grasses, is shown in the table—which is based on the results of Way's analyses:—

COMPOSITION OF THE HAY OF ARTIFICIAL GRASSES.

	Flesh-forming Substances.	Fatty Matters.	Respiratory Substances.	Woody fibre.	Ash.	Water.
Trifolium pratense—red clover	18·79	3·06	37·06	16·46	7·97	16·6
Trifolium pratense perenne—Purple clover	15·98	3·41	35·35	21·63	6·96	,,
Trifolium incarnatum—Crimson clover	13·83	3·11	31·25	26·99	8·15	,,
Trifolium medium—Cowgrass	20·27	2·97	30·30	20·12	9·67	,,
Do., second specimen ...	15·64	3·98	41·38	15·70	6·64	,,
Trifolium procumbens—Hop trefoil	17·07	3·89	36·55	18·88	6·94	,,
Trifolium repens—White trefoil	15·63	3·65	33·37	22·11	8·57	,,
Vicia sativa — Common Vetch	19·68	2·55	32·87	22·82	5·42	,,
Vicia sepium — Bush vetch	19·23	2·40	27·62	25·87	8·21	,,
Onobrychis sativa — Sainfoin	15·38	2·51	38·30	20·59	6·56	,,
Medicago sativa — Lucerne	10·63	2·30	33·47	28·51	8·42	,,
Medicago lupulina — Yellow clover	20·50	3·38	27·76	22·66	9·03	,,
Plantago lanceolata — Rib grass	11·91	3·06	33·58	27·56	7·23	,,
Poterium sanguisorba—Burnet	13·96	3·34	39·50	19·89	6·64	,,
Achillea millefolium—Millefoil	8·62	2·09	37·88	27·24	7·50	,,
Mean	15·81	3·18	34·42	22·47	7·59	16·6

Very many analyses of hay have been made by British and Continental chemists, the results of which are of great interest to the agriculturist. The composition of the natural and artificial grasses, which is shown in the tables given in pages 158-9 will, if we reduce their per-centage of water to 16, give us an approximation to the composition of hay. If the herbage, too, be sown in the proper time, and the hay-making process be skilfully conducted, there will be but little difference, except in the amount of water, between the plants in their fresh and dry state; but owing to inopportune wet weather, and carelessness in manipulation, excellent herbage is not unfrequently converted into inferior hay.

According to Dr. Voelcker, the average composition of meadow-hay, as deduced from the results of twenty-five analyses, is as follows:—

Water	14·61
Flesh-forming constituents	8·44
Respiratory and fatty matters	43·63
Woody fibre	27·16
Mineral matter (ash)	6·16
	100·00

Dr. Anderson's analysis of meadow-hay, one year old, and of inferior quality, gave the following results:—

Water	13·13
Flesh-forming matters	4·00
Non-nitrogenous substances	77·61
Mineral matter	5·26
	100·00

The results of the investigations of Way prove that the herbage of water-grass meadows is more nutritious than that of dry meadows—results perfectly harmonious with the experience of practical men.

It is a somewhat general belief, that the aftermath, or second cutting, is less nutritious than the first cutting; but there appears to be no chemical difference between the two crops, provided they be saved under equally favorable conditions. According to Dr. Anderson, the composition of clover-hay of the second cutting is as follows :—

Water	16·84
Flesh-forming principles ...	13·52
Non-nitrogenous matters ...	64·43
Mineral matter (ash) ...	5·21
	100·00

I have already shown the importance of reaping in proper season—not less necessary is it to mow before the plants ripen fully, and even before they flower. The results of the experiments of Stöckhardt, Hellreigel, and Wolff, in relation to this point, are very interesting, and are well worthy of re-production here.

RESULTS OF STÖCKHARDT'S AND HELLREIGEL'S EXPERIMENTS.

	Stem.			Leaves.		
	Water in Fresh Plant.	Hay.		Water in Fresh Plant.	Hay.	
		Flesh-forming Matters.	Ash.		Flesh-forming Matters.	Ash.
Clover cut on the						
4th June, quite young...	82·80	13·16	9·71	83·50	27·17	9·42
23rd ,, ready for cutting	81·72	12·72	9·00	82·68	27·69	9·00
9th July, beginning to flower	82·41	12·40	6·12	77·77	15·83	10·46
29th July, full flower ...	78·30	9·28	4·63	70·80	19·20	9·58
21st August, ripe... ...	69·40	6·75	4·82	65·70	18·94	12·33

RESULTS OF WOLFF'S EXPERIMENT.

	Red Clover.				Alsike Clover.			
	Beginning to flower, 11th June.		Full flower, 25th June.		Beginning to flower, 23rd June.		Full flower, 29th June.	
	Fresh.	Hay.	Fresh.	Hay.	Fresh.	Hay.	Fresh.	Hay.
	pr.cent.	pr.cent.	pr.cent.	pr.cent.	pr.cent.	pr.cent.	pr.cent.	pr.cent.
Water	83·07	16·66	76·41	10·66	86·98	16·66	82·60	16·66
Ash	1·43	7·04	1·67	5·90	1·12	7·17	1·45	6·94
Woody fibre ...	4·24	20·87	8·88	37·37	3·79	24·26	5·11	24·47
Nutritive sub-stances ...	11·26	55·43	13·04	46·07	8·11	51·91	10·84	51·93

During the operation of converting the grass—"natural" or "artificial"—into hay, there is more or less loss of nutritive matter sustained by fermentation, the dispersion of the smaller leaves by the wind, and other agencies. But this unavoidable loss is trivial when compared with the prodigious waste sustained, in Ireland at least, by allowing the hay to remain too long in cocks in the field. "Within the last three or four years," says Mr. Baldwin, of the Glasnevin Albert Model Farm, "we have made agricultural tours through twenty-five of the thirty-two counties of Ireland; and from careful consideration of the subject, and having in some instances used a tape-line and weighing-machine to assist our judgment, we have come to the conclusion that one-twentieth of the hay-crop of Ireland is permitted to rot in field-cocks. The portion on the ground, as well as that on the outside of the cocks, is too often only fit for manure. And the loss of aftermath, and of the subsequent year's crop (if hay or pasture), suffers to the extent of from sixpence to one shilling per acre. If we unite all these sources, the loss sustained annually in this country is something serious to contemplate. On an average, for all Ireland, it is not under 20 per cent., or a fifth of the actual value of the crop." This is a startling statement; but I do not believe it to be an exaggeration of the actual state of things.

Damaged Hay and Straw.—Damaged corn and potatoes, so much injured as to be unfit for human food, are generally given, and with apparently good results, to the inferior animals. The "meat manufacturing machines," as the edible varieties of the domesticated animals are now generally termed, are not very dainty in their choice of food; and vegetable substances which would excite the disgust of the lords of the creation are rendered nutritious and agreeable by being reorganised in the mechanisms of oxen, sheep, and pigs.

Now, although it is pretty generally known that musty corn and diseased potatoes form good feeding stuffs, it is not so patent whether or not the natural food of stock, such as hay and straw in a diseased state, is proper food for those animals.

This question is worthy of consideration. Firstly, I shall describe the nature of the diseases which most frequently affect fodder; these are, "mildew" and "mould." These diseases are produced by the ravages of minute and very low forms of vegetable life, termed by the botanists *epiphytical fungi*. The mildew (*Puccinia graminis*) generally attacks the grasses when they are growing, and is more frequently met with on rich and heavily manured soils. In localities where heavy night-fogs and dews are of common occurrence, this pest often destroys whole crops. On the other hand, in light, sandy, and well-drained soils, and in warm and dry districts, the mildew is a rare visitant. The "blue mould" (*Aspergillis glaucus*) attacks hay and straw in the stack or rick, and without any regard to their origin—no matter whether they were the produce of the wettest or the dryest, the warmest or the coldest of soils. The chief condition in the existence of the blue mould is excessive moisture. If the hay or straw be too green and succulent when put up, or if rain get at them in the rick, the mould is very likely to make its appearance, and the well-known odor termed *musty* will speedily be developed.

Neither the mildew nor the mould can, strictly speaking, be regarded as parasites, such as, for example, the flax-dodder, which feeds upon the healthy juices of the plant to which it is attached. It appears to me that the tissues and juices of the fodder-plants decay *first*, and then the mould or the mildew appears and feeds upon the decomposing matter. Now, as these vegetables belong to a poisonous class of fungi, it is more than probable that they convert the decomposing substance of the straw or hay into unwholesome, if not poisonous matter ; and it is not unlikely but that the disagreeable odor which they evolve is designed by nature as a sign to the lower animals not to partake of mouldy food. There is no doubt but that most animals will instinctively reject fodder in this state ; and the question arises, ought this odour to be destroyed or disguised. in order to induce the animals to eat the damaged stuff? The experience of most feeders who have largely consumed mouldy provender is, that although cattle may be induced to eat it, they never thrive upon such stuff if it form a heavy item in their diet The reason of this is obvious. The nitrogenous portion of the straw is that which is chiefly assimilated by the fungi. And as this constituent is the one which contributes to the formation of muscle, and is naturally extremely deficient in straw and hay—more particularly the former—it follows that the animals fed upon mouldy fodder cannot elaborate it into lean flesh (muscle).

In the case of young stock, mouldy fodder is altogether inadmissible, for these animals require abundance of flesh-forming materials—precisely those which the fungi almost completely remove from the diseased fodder.

As large quantities of mouldy or mildewed provender are at the present moment to be found in many farmsteads, and as they are unsaleable, and must therefore be made use of in some way at home, it is well to consider the best way to dispose of them. In the case of straw, the greater portion will be required for litter, and if the whole of the damaged article can be disposed of in this way so much the better. If, however, there is more than is necessary for the bedding of the stock, it may be used in conjunction with sound fodder, but always in a cooked state. The greater part, if not the whole, of the diseased nitrogenous part of the straw is soluble in warm water, so that if the fodder be well steamed the poisonous matter will be eliminated to such an extent as to leave the article almost as wholesome as good straw, but not so nutritious. The straw cleansed in this way will be very deficient in flesh-forming, though not in fat-forming power, and this fact should be duly considered when the other items of the animal's food are being weighed out. Beans, malt-combs, and linseed-cake are rich in muscle-forming principles, and are consequently suitable adjuncts to damaged fodder; but the latter should never constitute the staple food, or be given unmixed with some sweet provender.

When the fodder is considerably damaged it becomes,

after steaming, nearly as tasteless as sawdust. To this kind of stuff the addition of a small amount of some flavorous material is very useful. For damaged hay, Mr. Bowick recommends the following mixture:—

Fenugreek (powdered)	...	...	112 parts.
Pimento	...	...	...	...	4 ,,
Aniseed	...	...	...	...	4 ,,
Caraways	...	...	...	...	4 ,,
Cummin	...	...	...	...	2 ,,

A pinch of this compound will render agreeably-flavored the most insipid kinds of fodder.

Mr. Bowick states that he had fed large numbers of bullocks on damaged hay, flavored with this compound, and that their health was not thereby injured in the slightest degree.

SECTION V.

ROOTS AND TUBERS.

THE important part which the so-called root crops play in the modern systems of agriculture, has secured for them a large share of the attention of the chemist, so that our knowledge of their composition and relative nutritive value is very extensive. As compared with most other articles of food, the roots, as they are popularly called, of potatoes, turnips, mangels, carrots, and such like plants, contain a high proportion of water, and are not very nutritious; indeed, with the exception of the potato, none of them contain 20 per cent. of solid matter, and some not more than five per cent. They are, however, easily produced in great quantities, which compensates for their low nutritive value. I shall consider each of the more important roots separately.

The Turnip.—There are numerous varieties of this plant, which differ from each other in the relative proportions and total amount of their constituents, and even in different individuals of the same variety there is considerable variation in composition; hence the difficulty which has been felt by those who have endeavored to assign to this plant its relative nutritive value. From the average results of a great number of experiments, conducted both in the laboratory and the feeding-house, it is concluded that turnips are the most inferior roots produced in the field. The Swedish turnips are the most valuable kind: they contain a higher proportion of solid matter than the other varieties, and they are firmer and store better. The average composition of five varieties of turnips, as deduced from the results of the analyses of Anderson and Voelcker, is shown in the following table:—

ANALYSES OF TURNIPS.

	Swedish Turnip.	White Globe.	Aberdeen Yellows.	Purpletop Yellows.	Norfolk Bell.
Water	89·460	90·430	90·578	91·200	92·280
Albuminous, or flesh-forming substances...	1·443	1·143	1·802	1·117	1·737
Non-nitrogenous, or fat-forming substances (fat, gum, sugar, &c.)	5·932	5·457	4·622	4·436	2·962
Woody fibre	2·542	2·342	2·349	2·607	2·000
Mineral matter (ash) ...	0·623	0·628	0·649	0·640	1·021
	100·000	100·000	100·000	100·000	100·000

The *Greystone Turnip* is a variety which has only quite recently been introduced. It is stated to be an uncommonly productive crop, usually yielding returns from 30 to 50 per cent. greater than those obtained from other varieties of the turnip. The composition of the Greystone turnip appears to be inferior, so that probably it is not, after all, a more economical plant than the ordinary kinds of turnips.

DR. ANDERSON'S ANALYSIS OF THE GREYSTONE TURNIP.

	No. 1. Grown on Clay.	No. 2. Grown on Sand.
Water	93·84	94·12
Oil	0·26	0·34
Soluble albuminous matters	0·35	0·56
Insoluble ditto	0·20	0·18
Soluble respiratory matters	2·99	2·32
Insoluble ditto (chiefly fibre)	1·73	1·85
Ash	0·63	0·63
	100·00	100·00

It was at one time the fashion—not yet become quite obsolete—to regard the proportion of nitrogen in the turnip as the measure of the nutritive value of the bulb; but the fallacy of this opinion has been shown by several late investigators, and more particularly by the results of one of the numerous series of feeding experiments conducted by Mr. Lawes. Many bulbs exceedingly rich in nitrogen are very deficient in nutritive power—partly from a deficiency in the other elements of nutrition—partly because most of their nitrogen is in so low a degree of elaboration as to be incapable of assimilation by animals. The value of a food-substance does not merely depend upon the amount and the relative proportion of its constituents, but also, and to a very great extent, upon their easy assimilability. There is but little doubt that the nutritive matters contained in the Swedish turnip when the bulb is fresh are very crude. By storing, certain chemical changes take place in the bulb, which render it more nutritious and palatable. A large proportion of the non-nitrogenous matters exist in the fresh root as pectin; but this substance, if the bulb be preserved for a couple of months, becomes in great part converted into sugar, which is one of the most palatable and fattening ingredients of cattle-food. By storing, too, the bulbs lose a portion of their excessive amount of water, and become less bulky, which is unquestionably a desideratum. These facts suggest the necessity for cultivating the earlier varieties of the turnip, for it may be fairly doubted if a late-grown crop, left for consumption in the field, ever, even under the most favorable circumstances, attains its perfect development. At the same time it must not be forgotten that turnips *fully matured* in the field rather deteriorate than otherwise after a few weeks' storage.

Many agriculturists consider that there is a strict relation between the specific gravity, or comparative weight of the bulb, and its nutritive value; others believe that a very large turnip must necessarily be inferior in feeding qualities to a small one; whilst not a few maintain that neither its size nor its specific gravity is an indication of its feeding qualities. Dr. Anderson, who has specially investigated a portion of this subject, states that "the specific gravity of the whole turnip cannot be accepted as indicating its real nutritive value, the proportion of air in the cells being the determining element in such results; that there is no constant relation between the specific gravity of, and the nitrogen compounds in, the bulb; and that such relation does exist between the specific gravity of the expressed juice and the nitrogen compounds and solid constituents." Dr. Anderson allows, however, that the best varieties of the turnip have the highest specific gravity; which admission—coupled with the fact admitted by all experimenters that the heavy roots store best—lead me to adopt the opinions of those who consider great specific gravity as one of the favorable indications of its nutritive value. With respect to size, I prefer bulbs of moderate dimensions; the monsters that win the prizes at our agricultural shows—and which, in general, are *forced*—are inferior in feeding qualities, are always *spongy*, and almost invariably rot when stored.

The composition of the turnip is influenced not only by the nature of the soil on which it is grown, but also by that of the manure applied to it. The most reliable authorities are agreed that turnips raised on Peruvian guano are watery, and do not keep well; but that with a mixture of Peruvian guano and superphosphate of lime, with phospho-guano, or with farmyard manure supplemented with a moderate amount of guano, the most nutritious and firm bulbs are produced.

Turnip-tops have been analysed by Voelcker, with the following results :—

ONE HUNDRED PARTS CONTAIN—

	White.	Swedish.
Water	91·284	88·367
Nitrogen compounds	2·456	2·087
Non-nitrogenous matters (gum, sugar, &c.)	0·648	1·612
Ditto, as woody fibre	4·092	5·638
Mineral matter...	1·520	2·296
	100·00	100·000

These figures apparently show that the tops of turnips are more valuable than their bulbs; but, in the absence of any feeding experiments made to determine the point, we believe they are less so, as a very large proportion of the solid matter in the tops of turnips is in too low a degree of elaboration to be assimilable. Their high proportions of nitrogen and mineral matter constitute them, however, a very useful manure—nearly twice as valuable as the bulbs; this fact should be borne in mind when turnips are sold off the land.

The Mangel-wurtzel is one of the most valuable of our green crops. Its root is more nutritious than the turnip, occupying a position in the scale of food equivalents midway between that bulb and the parsnip. Mangels, when fresh, possess a somewhat acrid taste, and act as a laxative when given to stock; but after a few months' storing they become sweet and palatable, and their *scouring* property completely disappears.

Although the mangel is one of the most nutritious articles of food which can be given to cattle, yet it is stated on the best authority that sheep do not thrive upon it. Voelcker, who has investigated this subject, informs us that a lot of sheep which he fed on a limited quantity of hay and an unlimited quantity of mangels, did not, during a period of four months, increase in weight, whilst another lot of sheep supplied with a small quantity of hay, and Swedish turnips *ad libitum* increased on an average 2½ lbs. weekly. I believe the experience of the greater number of feeders agrees with the results of Dr. Voelcker's experiment.

The chemistry of the mangel-wurtzel has been thoroughly studied by Way and Ogston, Fromberg, Wolff, Anderson, and Voelcker. According to the last-named chemist, its average composition is as follows :—

Water	87·78
Flesh-forming matters	1·54
Sugar	6·10
Gum, pectin, &c.	2·50
Woody fibre	1·12
Mineral matter (ash)	0·96
	100·00

It is difficult to accurately determine by a comparative trial the relative feeding properties of mangels and turnips, for the former are only in a fit state to be given to the animals when the latter are deteriorating. However, by comparing the composition of the two substances, and the results obtained from numerous feeding experiments, it would appear, that on the average 75 lbs. weight of mangels are equal to 100 lbs. weight of turnips. Of the different varieties of the mangel the long yellow appears to be the most nutritious, and the long red the least so.

The leaves of the mangel—some of which are occasionally pulled and used for feeding purposes, during the growth of the bulb—are an excellent feeding substance : their composition indicates a nutritive value but little inferior to that of the root ; but as their constituents cannot be in a highly elaborated condition, it is probable they are not more than equal to half their weight of the bulbs.

One *questio vexata* of the many which at present occupy the attention of the agricultural world is, whether or not the leaves of mangels may be removed with advantage during the latter part of the development of the plants. This practice prevailed rather extensively a few years since, but latterly it has fallen somewhat into disuse.

Those who adopt this plan urge, as its advantages, that a large quantity of food is obtained at a time when it is urgently needed, and that instead of the removal of the leaves exercising an injurious influence on the development of the roots, the latter are actually increased in size.

In 1859 an experimental investigation was carried out at the Glasnevin Model Farm, with the view of throwing new light on the question. The outside leaves were very gradually removed on different occasions—from the 12th August to the 15th October. In this way five tons of leaves per statute acre were removed, and subsequently made use of for feeding purposes. The experiment was conducted on a field of four acres, of which the produce of 12 drills, each 200 yards in length, was left untouched. The result was that the produce of the roots of the untouched plants was only 40 tons 8 cwt. 6 qrs. per acre, whilst the roots of the plants which had been partly denuded of their leaves weighed at the rate of 45 tons 1 cwt. This experiment afforded results which are apparently favorable to the practice of stripping the leaves; but it is to be regretted that it was not rendered more complete by an analysis of the roots, as a great bulk of roots does not necessarily imply a great weight of dry food, and it is just possible, though not very probable, that the roots of the stripped mangels contained a larger proportion of water than those of the untouched plants.

The results of the experiments of Buckman, and of Professor Wolff, of the Royal Agricultural College at Hohenheim, are at direct variance with those obtained at Glasnevin. Both of these experimenters found that the removal of the leaves occasioned a diminution in the produce of the roots to the amount of 20 per cent. Nor was this the only loss, for it was found by the German professor that the roots of the untouched plants possessed a far higher nutritive value than those of the stripped mangels.

When doctors differ, who is to decide? Here we have high authorities in the agricultural world at direct variance on a matter of fact. The names of Buckman and Wolff are a sufficient guarantee that the experimental results which they announce are trustworthy, and I can testify, from observation, that no field experiments could be more carefully conducted than those carried out at the Albert Model Farm. We can only, then, under the circumstances, admit that both Mr. Boyle, on the one side, and Professors Buckman and Wolff on the other, are correct in their statements of fact; but as it is evident both cannot be right in the general inferences therefrom, it is desirable that the subject should be still further investigated, and the truth be placed beyond doubt. It is a question which appears so simple that one is at a loss to account for the discrepant opinions in relation to it which prevail. "Let nothing induce the growers," says Mr. Paget, in a paper on the cultivation of the mangel, "to strip the leaves from the plant before taking up the root. A series of careful experiments has convinced me that by so doing we borrow food at a most usurious interest." "Although," says Mr. Boyle, "the practice of stripping has been followed for many years on the farm without any perceptible injury to the crop, these results, showing so considerable an addition to the crop from taking off the leaves, were hardly anticipated." It certainly does appear somewhat at variance with our notion of the functions of the leaves of plants, that their partial removal could possibly cause an increase in the weight of the roots; but granting such to be the fact, it is not altogether *theoretically* inexplicable. We know that highly nitrogenous manure has a tendency to increase the development of the leaves of turnips at the *expense* of the roots. Gardeners, too, not unfrequently remove some of the buds from their fruit trees, lest the excessive development of foliage should retard or check the *growth* of the fruit. *Theoretically* an excessive development of the leaves of the mangel may be inimical to the growth of the root. Probably, too, it may be urged, the outer leaves, which soon become partially disorganised and incapable of elaborating mineral matter into vegetable products, prevent the access of light to the more vigorous inner leaves. In conclusion, I may say of this subject that it is worthy of further elucidation; and I would suggest to my readers, and more especially to the managers of the various model farms, the desirability of fully testing the matter.

The *White Beet* is a congener of the mangel. It is largely grown on the continent as a sugar-producing plant, but is seldom cultivated in these countries. It produces about 15 tons of roots per acre, and its roots on the average contain—

Water	83·0
Sugar	10·0
Flesh-formers	2·5
Fat-formers	1·5
Fibre	2·0
Ash	1·0
	100·0

This plant is deserving of more extensive growth in Great Britain.

The *Parsnip* is, after the potato, the most valuable of roots. It differs from the turnip and the mangel in containing a high proportion of starch, and but little sugar; and its flesh-forming constituents are largely made up of casein, instead of, as in the case of the turnip, albumen.

The average composition of the parsnip is as follows:—

Water	82·00
Flesh-forming principles	1·30
Fat-formers (starch, sugar, &c.)	7·75
Woody fibre	8·00
Mineral matter (ash)	0·95
	100·00

The parsnip is extensively grown in many foreign countries, on account of its valuable feeding properties. As a field-crop it is but little cultivated in Great Britain, and its use is—if we except the table—almost restricted to pigs. Its food equivalent is about double that of the turnip; that is, one pound of parsnips is equal to two pounds of turnips.

The *Carrot* bears a close resemblance to the parsnip, from which, however, it differs, containing no starch, and being somewhat inferior in nutritive value. According to Voelcker, its average composition is as follows :—

Water	88·50
Flesh-formers	0·60
Fat-formers (including woody fibre)	10·18
Mineral matter (ash)	0·72
	100·00

As carrots contain a high proportion of fat-forming matters, and a low per-centage of flesh-forming substances, they are better adapted for fattening purposes. Dairy stock greedily eat them; and they are given with great advantage to horses out of condition.

Kohl-Rabi.—This plant, though early introduced into the agriculture of these countries, has made but little progress in the estimation of the farmer. It belongs to the order and genus which include the turnip, but differs widely from that plant in its mode of growth. Its bulb—which is formed by an enormous development of the overground stem—is, according to some authorities, less liable than the turnip to injury from frost. It is subject to no diseases, save anbury and clubbing; and, owing to its position above the soil, it can be readily eaten off by sheep. The bulbs store better than Swedes, and, according to some farmers, keep even better than mangels. With respect to the flavor of this bulb, there is some difference of opinion. Professor Wilson, of Edinburgh, quotes several eminent feeders to prove that "whether in the fold for sheep, in the yard for cattle, or in the stables for horses, it will generally be preferred to the other descriptions of home-grown keep." Mr. Baldwin, on the contrary, states that although good food for sheep, it is too hard-fleshed for old ewes, and that carrots are better food for horses, and Swedish turnips for cattle.

An accurately conducted comparative trial to test the nutritive value of the Kohl-rabi, was conducted at the Glasnevin Model Farm, under the direction of Mr. Baldwin. The experiment was commenced in January, 1863. Four oxen were selected, and divided into two lots. Nos. 1 and 2 (Lot 1) were fed on Kohl-rabi, oil-cake, and hay, and Nos. 3 and 4 (Lot 2) on Swedish turnips, oil-cake, and hay. As the animals supplied with the Kohl-rabi did not appear to relish it, and as it was desirable to gradually accustom them to the change of food, the experiment did not really commence till the 12th January. On that date the weights of the animals were as follows :—

		cwt.	st.			cwt.	st.
Lot 1.	No. 1.	10	1	Lot 2.	No. 3.	7	5
	No. 2.	7	4		No. 4.	10	2
		17	5			17	7

The lots, therefore, counterpoised each other pretty fairly. From the 12th to the 28th January they received the following quantities of food per diem :—

		1.	2.	3.	4.
Roots ...	stones	$7\frac{1}{2}$	6	6	$7\frac{1}{2}$
Oil cake	pounds	$4\frac{1}{2}$	3	3	$4\frac{1}{2}$
Hay ...	pounds	$10\frac{1}{2}$	$10\frac{1}{2}$	$10\frac{1}{2}$	$10\frac{1}{2}$

The animals fed upon the Kohl-rabi evinced from the first a disinclination to it, but they nevertheless ate it before their meal of oil-cake was supplied to them. On the morning of the 28th January they were put upon the dietary shown in the table, and which induced them to eat the Kohl-rabi more quickly.

		1.	2.	3.	4.
At 6.30 a.m.	Roots, Stones ...	3	$2\frac{1}{2}$	$2\frac{1}{2}$	$3\frac{1}{2}$
	Cake, lbs.	$1\frac{1}{2}$	1	1	1
At 12.30 a.m.	Roots, Stones ...	3	$2\frac{1}{2}$	$2\frac{1}{2}$	$3\frac{1}{2}$
	Cake, lbs.	$1\frac{1}{2}$	1	1	1
At 6.30 p.m.	Roots, Stones ...	3	$2\frac{1}{2}$	$2\frac{1}{2}$	$3\frac{1}{2}$
	Cake, lbs.	$1\frac{1}{2}$	1	1	1
At 9.30 p.m.	Hay, lbs.	7	7	7	7

On the 11th February the cattle were again weighed, when their increase was found to be as follows:—

		Weight on Jan. 12. cwt. st.	Weight on Feb. 11. cwt. st.	Increase in 30 days. st.
1 }	Lot 2, fed on Kohl- }	10 1	10 4	3
2 }	rabi, &c.... }	7 4	7 6	2
	Total ...	...	...	5
3 }	Lot 2, fed on Swedes, }	7 5	8 3	6
4 }	&c. ... }	10 2	10 7¼	5½
	Total ...	...	...	11½

The results of this experiment show that the animals fed upon Swedish turnips, hay, and oil-cake, increased in weight at a rate more than 100 per cent. greater than the lot supplied with equal quantities of Kohl-rabi, hay, and oil-cake. The superiority of the Swedish turnips was rendered more evident by the results of subsequent experiments. Nos. 1 and 4 were not tried after the 11th February; but Nos. 2 and 3 were kept under experiment. No. 2 was put on Swedes, and No. 3 on mangel-wurtzel, and after an interval of a fortnight No. 2 had increased much more than they had done on Kohl-rabi.

Specimens of the Kohl-rabi and Swedish turnips employed in this experiment were submitted to me for analysis by Mr. Baldwin, and yielded the following results:—

	Kohl-rabi.	Swedish Turnip.
Water	87·62	88·84
Nitrogenous, or flesh-forming principles	2·24	1·66
Non-nitrogenous, or fat-forming principles	7·78	6·07
Woody fibre	1·34	2·73
Mineral matter (ash)	1.22	0·70
	100·00	100·00

These results show a slight superiority of the Kohl-rabi over the Swedish turnip; the great difference in their nutritive power, as shown by Mr. Baldwin's experimental results, must

O

therefore be due to the superior flavor and digestibility of the turnip.

Dr. Anderson's analysis of Kohl-rabi afforded results more favorable to the highly nutritive character assigned by some feeders to that bulb than those arrived at by me. The bulbs, it should however be remarked, were grown, no doubt with great care, by Messrs. Lawson and Son, the well-known seedsmen:—

ANALYSIS OF KOHL-RABI, BY DR. ANDERSON.

	Bulbs.	Tops.
Water	86·74	86·68
Flesh-forming principles	2·75	2·37
Fat-forming principles	8·62	8·29
Woody fibre	0·77	1·21
Mineral matter	1·12	1·45
	100·00	100·00

The *Radish* is a plant which deserves a place amongst our field crops, though hitherto its cultivation has been restricted to the garden. At one time its leaves were boiled and eaten, but in these latter days they are subjected to neither of these processes. The root, however, in its raw state, is, as every one is aware, considered one of the dainties of the table.

Many of those who devote themselves to the important study of dietetics, consider the use of raw vegetables to be objectionable; but be their objections groundless, or the reverse, it is certain that a vegetable which, like the radish, may be eaten raw with apparently good results, cannot be otherwise than a good article of food when cooked. I once tried the experiment of eating matured radishes, not as a salad, but cooked like any other boiled vegetable, and I must say that I found their flavor rather agreeable than otherwise. Boiled radishes—roots and tops—form excellent feeding for pigs. How could it be otherwise? for what is good for the family of man must surely be a luxury to the swine tribe. I have known horses to eat radishes greedily, and I am certain that they would prove acceptable to all the animals of the farm. But it may be asked, why it is that I recommend the use of radishes as food for stock, when there are already so many more nutritious roots at our disposal—turnips, mangels, and potatoes. Simply for this reason:—Between the departure of the roots and the advent of the grasses, there is a kind of interregnum.* Now we want a good tuberous, bulbous, or tap-rooted plant to fill up this interregnum. Such a plant we have in the radish. The root is certainly a small one, but then it grows so rapidly that a good supply can be had within thirty days from the sowing of the seed, and a crop can be matured before the time for sowing turnips. Two crops may be easily obtained from land under potatoes—one before the tops cover the ground, the other after the tubers have been dug out. The yield of radishes, judging from the produce in the garden, would be at least six tons of roots and three tons of tops. I would suggest, then, that the radish should at once get a fair chance as a stolen crop. If it succeed as such, it will not be the first gift of the gardener to the husbandman. Was not the mangel-wurtzel once known only as the produce of the garden?

The composition of the radish indicates a nutritive value less than that of the white turnip. I have analysed both the root and the tops, and obtained the following results:—

ANALYSIS OF THE RADISH.

	Root.	Tops.
Water	95·09	94·30
Flesh-forming principles	0·52	0·75
Fat-formers (starch, gum, fat, &c.)	1·06	1·16
Woody fibre	2·22	2·36
Mineral matter (ash)	1·11	1·43
	100·00	100·00

The *Jerusalem Artichoke* has long been cultivated as a field-crop on the Continent, and in certain localities the breadth

* Unless when Kohl-rabi is cultivated, for the bulbs of this plant may be preserved in good condition up to June. I have advocated the cultivation of the radish as a food crop in the "Agricultural Review" for 1861.

occupied by it is very considerable. The French term the tuberous root of this plant *poitre de terre*, or *topin ambour;* and although they expose it for sale in the markets, it is not much relished by our lively neighbours, who are so remarkable for their *cuisiniere*. As food for cattle, however, the French agricultural writers state it to be excellent. It is much relished by horses, dairy cows, and pigs; store horned-stock also eat it when seasoned with a little salt, and appear to enjoy it amazingly when permitted to pull up the roots from the soil. The green tops are also given to sheep and cattle, and, it is stated, are readily eaten by those animals.

The Jerusalem artichoke (*Helianthus Tuberoses*) differs from its half namesake, the common artichoke, and resembles the potato in being valuable chiefly for its tubers. It is perennial, and attains on the Continent a height varying from 7 to 10 feet. In this country its dimensions are less. The stem is erect, thick, coarse, and covered with hairs. It is a native of Mexico, and although introduced 200 years ago into Europe, it can hardly be said to be acclimatised, since it very seldom flowers, and never develops seed. The plant is therefore propagated by cuttings from its tubers, each containing one or two eyes; or if the tubers be very small, which is often the case, a whole one is planted. The tubers possess great vitality, and remain in the ground during the most severe frosts, without sustaining the slightest injury. For this reason it is usual to devote a corner of the garden to the cultivation of the Jerusalem artichoke; for, no matter how completely the crop may appear to have been removed from the soil, portions of the tubers will remain and shoot up into plants during the following season. This peculiarity of the plant it is likely may prove an obstacle to its having a place assigned to it in the rotation system.

The question now presents itself—What are the peculiar advantages which the crop possesses which should commend it to the notice of the British farmer? I shall try to answer the question.

1st. No green crop (except furze) can be grown in so great a variety of soils; except marshy or wet lands, there is no soil in which it refuses to grow.

2nd. It does not suffer from disease, is very little affected by the ravages of insects, is completely beyond the influence of cold, and may remain either above or below ground for a long time without undergoing any injurious changes in composition.

3rd. It gives a good return, when we consider that it requires very little manure, and but little labor in its management.

At Bechelbronn, the farm of the celebrated Boussingault, the average yield is nearly eleven tons per acre, but occasionally over fourteen tons is obtained. Donoil, a farmer of Bailiere, in the department of Haut-loire, states that he fed sheep exclusively on the tops and tubers of this plant, and that he estimated his profits at £23 per hectare (£9 3s. 4d. per acre). The soil was very inferior. Donoil terms it third-rate, and it does not appear to have been manured even once during the fifteen years it was under Jerusalem artichoke. I fear our artificial manure manufacturers will hardly look with a favorable eye on the advent of a crop into our agriculture which can get on so well without the intervention of any fertilising agents. Indeed, several of the French writers state that little or no manure is necessary for this plant. But this can hardly be the case; for it is evident that a crop which, according to Way and Ogston, removes 35 lbs. of mineral matter per ton from the soil, or three times as much potash as turnips do, must certainly be greatly benefited by the application of manure. And I have no doubt but that the Jerusalem artichoke, if well manured and grown in moderately fertile soil, would produce a much heavier crop than our Continental neighbors appear to get from it.

4th. The Jerusalem artichoke may be cultivated with advantage in places where ordinary root-crops either fail or thrive badly. In such cases the ground should be permanently devoted to this crop. Kade gives an instance where a piece of indifferent ground had for thirty-three years produced heavy crops of this plant, although during that time neither manure nor labor had been applied to it. In Ireland the potato has been grown under similar circumstances.

The nutritive constituents of tubers of the Jerusalem artichoke bear a close resemblance in every respect, save one, to those of the potato. Both contain about 75 per cent. of water, about 2 per cent. of flesh-forming substances, and 20 per cent. of non-nitrogenous, or fat-forming and heat-giving elements. In one respect there is a great difference—namely, that sugar makes up from 8 to 12 per cent. of the Jerusalem artichoke, whilst there is but a small proportion of that substance in the potato.

The large quantity of sugar contained in this root is no doubt the cause of its remarkable keeping properties in winter, and it also readily accounts for the avidity with which most of the domesticated animals eat it.

On the whole, then, I think that the facts I have brought forward relative to the advantages which the Jerusalem artichoke presents as a farm crop, justify the recommendation that it should get a fair trial from the British farmer, who is now so much interested in the production of suitable forage for stock.

COMPOSITION OF (DRY) JERUSALEM ARTICHOKE.

Albuminous matters	4·6
Fatty matters	0·4
Starch, gum, &c.	19·8
Sugar	69·5
Fibre and ash	5·7
	100·0

The *Potato*, regarded from every point of view, is by far the most important of the plants which are cultivated for the sake of their roots. Its tubers form the chief—almost sole— pabulum of many millions of men, enter more or less into the dietary of most civilised peoples, and constitute a large proportion of the food of the domesticated animals. The great importance of this plant, arising from its enormous consumption, has caused its composition to be very minutely studied by many British, Continental, and American chemists. With respect to its nutritive properties, the least favorable results were obtained by the American chemists, Hardy and Henry, and the most by the European chemists.

The flesh-forming principles vary from 1 per cent., as found by Hardy, to 2·41 per cent., the mean results of the analyses of Krocker and Horsford. The proportion of starch in different varieties of the potato also varies, but not to the same degree as the nitrogenous principles. In new potatoes, only 5 per cent. has been found; in ash-leaved kidneys, 9·50 per cent.; and in different kinds of cups, from 15 to 24 per cent. The amount of starch is also influenced by the soil, the manure, the climate, and the various other conditions under which the plant is developed. The proportion of starch increases during the growth, and diminishes during the storage of the tubers.

Dr. Anderson is the most recent investigator into the composition of the potato; the chief results of his inquiries are given in the following table :—

ANALYSIS OF THE POTATO BY DR. ANDERSON.

		Regents.	Dalmahoys.	Skerryblues.	White Rocks.	Orkney Reds.	Flukes.
Water	... ...	76·32	75·91	76·60	75·93	78·57	74·41
Starch	... ...	12·21	12·58	11·79	12·77	10·85	12·55
Sugar, &c.	... ...	2·75	2·93	3·09	2·17	2·78	2·89
Flesh-	soluble ...	2·16	2·10	1·90	1·88	1·48	1·98
formers	insoluble...	0·21	*0·15	0·16	0·24	0·21	0·20
Fibre	... ...	5·53	5·21	5·41	5·55	5·93	6·71
Ash	... ...	0·88	0·81	0·94	1·04	0·98	0·98
		100·06	99·69	99·89	99·58	100·80	99·72

The potato is relatively deficient in flesh-forming matters, and contains the respiratory elements in exceedingly high proportions; hence it is well adapted for fattening purposes, and in this respect is equal to double its weight of the best kind of turnips. When used as food for man, it should be supplemented by some more fatty or nitrogenous substance—such, for example, as flesh, oatmeal, or peas. Buttermilk, a fluid which is rich in nitrogen, is an excellent supplement to potatoes, and compensates to a great extent for the deficiency of those tubers in muscle-forming matters. If, then, the potato is destined to retain its place as the "national esculent" of the Irish, I trust their national beverage may be—so far at least as the masses of the people are concerned—buttermilk, and *not* whiskey.

Potatoes so far diseased as to be unsuited for use as food for man, may be given with advantage to stock. They may be used either in a raw or uncooked state, but the latter is the preferable form. Sheep do not like them at first, but on being deprived of turnips they acquire a taste for them; on a daily allowance, composed of 1 lb. of oil-cake or corn, and an unlimited quantity of potatoes, they fatten rapidly. Cattle thrive well on a diet composed of equal parts of turnips and diseased potatoes, and do not require oil-cake. The evening feed of horses may advantageously be composed of potatoes and turnips. If raw, the potatoes should be given in a very limited quantity—four or five pounds; in the cooked state, however, they may be given in abundance, but the animals should not, after their meal, be permitted to drink water for some hours. As a feeding substance, diseased potatoes, unless they be very much injured, are equal to twice their weight of white turnips; it is certain that they do not injure the health or impair the condition of the animals which feed upon them.

SECTION VI.

SEEDS.

In seeds the elements of nutrition exist not only in the most highly elaborated, but also in the most concentrated state; hence their nutritive value is greater than that of any other class of food substances.

Wheat Grain is the most valuable of seeds, as it contains, in admirably adjusted proportions, the bone, the fat, and the muscle-forming principles. In the form of bread, it has been, not inaptly, termed the "staff of life," for no other grain is so well adapted, *per se*, for the sustenance of man; and many millions of human beings subsist almost exclusively on it. The lower animals are in general fed upon the grain of oats, of barley, and of the leguminous plants, and the use of wheat is almost completely restricted to the human family.

Wheat grain, by the processes of grinding and sifting, is resolvable into two distinct parts—bran and flour. In twenty-four analyses made by Boussingault, the proportion of the bran was from 13·2 to 38·5 per cent., and that of the flour from 61·5 to 86·8 per cent. The floury part is of very complex structure; it includes starch, gluten, albumen, oil, gum, gummo-gelatinous matter, sugar,* and various saline matters. The gluten and albumen constitute the nitrogenous, or flesh-forming principles of flour, and make up from 16 to 20 per cent. of that substance; the non-nitrogenous, or fat-forming elements, such as starch and gum, form from 74 to 82 per cent. According to Payen, the proportion of gluten diminishes towards the centre of the seed, from which it follows that the part of the grain nearest the husk is the most nutritious—so far at least as muscle-making is concerned. The desire on the part of the public for very white bread has led to the *fine* dressing of Wheat-grain, and consequently to the separation from that substance of a very large proportion of one of its most nutritious constituents. Crude gluten may be obtained by kneading the dough of flour in a muslin bag under a small current of water; the starch, or fecula, and the gum, are carried away by the water, and the gluten in an impure form remains as an elastic viscous substance, which on drying becomes hard and brittle. It is to the gluten of flour that its

* According to some chemists, sugar does not exist in ripe grain, but is produced in it, during the process of analysis, by the action of the re-agents employed and the influence of the air.

property of panification, or bread-making, is due. On the addition of a ferment, a portion of the starch is converted into sugar and carbonic acid gas, and the latter causes the gluten to expand into the little cells, or vesicles, which confer upon baked bread its light, spongy texture.

ANALYSES OF WHEAT.

	1. Whole Grain.	2. Flour.	3. Bran.	4. Husk.
Water	15·00	14·0	13	13·9
Flesh-formers	12·00	11·0	14	14·9
Fat-formers	68·50	73·5	55	55·8
Woody fibre	2·75	0·7	12	9·7
Mineral matter	1·75	0·8	6	5·7
	100·00	100·0	100	100·0

Nos. 1, 2, and 3.—The mean results of a great number of analyses. No. 4.—By MILLON.

Over-ripening of Grain.—The final act of vegetation is the production of seed, after the performance of which function many plants, having accomplished their destined purpose, perish. The grasses (which include the cereals) are *annuals*, or plants which have but a year's existence, consequently their development ceases so soon as they have produced their seed. When wheat, oats, and the other cereals, attain to this final point in their growth, the circulation of their sap ceases, their color changes from green to yellow, and they undergo certain changes which destroy their power of assimilating mineral matter, and consequently render them no longer capable of increasing their weight.

The proper time for cutting wheat and the other cereals is immediately after their grain has been fully matured. When the green color of the straw just below the ears changes to yellow, the grain, be it ripe or unripe at the time, cannot afterwards be more fully developed. This is rendered impossible in consequence of the disorganisation of the upper part of the stem—indicated by, but not the result of, its altered hue—

which cuts off the supply of sap to the ears, and the latter do not possess the power of absorbing nutriment from the air.

When the vital processes which are incessantly going on in the growing plants are brought to a close, the purely chemical forces come into operation. If the seed be perfectly matured and allowed to remain ungathered, it is attacked in wet weather by the oxygen of the air, a portion of its carbon is burned off, some of its starch is converted into sugar, and in extreme cases it germinates and becomes *malty*. But not only is the seed liable to injury from the elements; it is also exposed to the ravages of the feathered tribe, and no matter how well a field of corn may be watched, or how great the number of *scarecrows* erected in it, there is always a certain diurnal loss, occasioned by the ravages of birds.

It is not only necessary that ripe corn should be cut as soon as possible, but it is sometimes desirable to reap it before it becomes fully matured. When the grain is intended for consumption as food, the less bran it contains the better. Now the bran, as is well known, forms the integument, or covering of the vital constituents of the seed; and it is the last part of the organ to be perfected. The growth of the seed for several days before its perfect development, is confined to the *testa* or covering. Now as this is the least valuable part of the article, its increase is matter of but little moment; and when it is excessive it renders the grain less valuable in the eyes of the miller. That the cutting of the grain before it is perfectly ripe is attended with a good result, is clearly proved by the results of an experiment recorded in Johnston's "Agricultural Chemistry." A crop of wheat was selected; one-third was cut twenty days before it was ripe; another third ten days afterwards; and the remaining portion when its grain had been fully matured. The relative produce in grain of the three portions taken, as stated above, was as 1, 1·325, and 1·260. The following table exhibits the relative proportions of their constituents :—

	In 100 parts of the grain cut at		
	20 days.	10 days.	Dead ripe.
Flour...	74·7	79·1	72·2
Sharps	7·2	5·5	11·0
Bran...	17·5	13·2	16·0
	99·4	97·8	99·2
The flour contained gluten ...	9·3	9·9	9·6

The results of this experiment, and of the general experience of intelligent growers, show that grain cut a week or ten days before it is perfectly ripe contains more flour, and of a better quality, too, than is found in either ripe or very unripe seed. But this is not the only advantage, for the straw of the green, or rather of the greenish-yellow corn, is fully twice as valuable for feeding purposes as that of the over-ripe cereals. There is an extraordinary decrease in the amount of the albuminous constituents of the stems of the cereals during the last two or three weeks of their maturation, and as there is not a corresponding increase of those materials in the seed, they must be evolved in some form or other from the plants.

There can be only one object attained by allowing the seed to fully ripen itself, and that is the insurance of its more perfect adaptability to the purpose of reproduction. When the *testa* is thick it best protects the germ of the future plant enclosed in it from the ordinary atmospheric influences until it is placed under the proper conditions for its germination.

Wheat, a costly food.—It occasionally happens that the wheat harvest is so abundant, that many feeders give large quantities of this grain to their stock. Now, as Indian corn is at least 25 per cent. cheaper than wheat, even when the price of the latter is at its *minimum*, I believe that it is always more economical to sell the wheat raised on the farm, and to purchase with the proceeds of its sale an equivalent of Indian corn, which is a more fattening kind of food.

Bran is, with perhaps the exception of malt-dust, the most nutritious of the refuse portions of grains. It is usually given to horses, and owing to its high proportion of nitrogen, is, perhaps, better expended in the bodies of those hard-working animals, than in those of pigs and cows—animals that occasionally come in for a share of this valuable feeding-stuff. It should be borne in mind that bran commonly acts as a slight laxative, and that it is less digestible than flour, a large portion of it usually passing through the animal's body unchanged. This drawback to the use of bran may be obviated by either cooking or fermenting the article, or by combining it with beans or some other kind of binding food.

AVERAGE ANALYSES OF GRAIN.

	Barley.	Bere.	Oats.	Oat-meal.	Indian Corn.	Rice.	Rye (Irish).	Buckwheat.
Water	16·0	14·25	14·0	13·00	14·5	14·0	16·0	14·19
Flesh-formers	10·5	10·10	11·5	16·00	10·0	5·3	9·0	8·58
Fat-formers	67·0	64·60	64·5	68·00	69·0	78·5	66·0	51·91
Woody fibre	3·5	9·03	7·0	1·75	5·0	2·5	8·0	23·12
Mineral matter	3·0	2·02	3·0	1·25	1·5	0·7	1·0	2·20
	100·0	100·00	100·0	100·00	100·0	100·0	100·0	100·00

Barley is inferior in composition to wheat. As a feeding stuff, the English farmers assign to it a higher, and the Scotch farmers a lower, place than oats, which, perhaps, merely proves that in Scotland the oat thrives better than the barley, and in England the barley better than the oat. Barley-meal is extensively used by the English feeders, and with excellent results. Where *barley-dust* can be obtained it is a far cheaper feeding stuff than the meal. Barley husks should never be given to animals unless in a cooked or fermented state.

Oat Grain is, perhaps, the most valuable of the concentrated foods which are given to fattening stock. When it is cheap it will be found a more economical feeding stuff than linseed-cake, and, unlike that substance, can be used without the fear of adulteration. Oats are equal to wheat in their amount of flesh-forming matters; but their very high proportion of indigestible woody fibre detracts from their nutritive value. Oat-meal is more nutritious than wheat-meal; and oat-flour, especially if finely dressed, greatly excels wheat-flour in its nutrimental properties, because, unlike the latter, the finer it is the greater is its amount of flesh-formers. Bread made of oat-flour is very heavy, and is far less palatable than the bread of wheat. Oat-meal has been found to contain nearly 20 per cent. of nitrogenous matters. The white oat is more nutritious than the black, and the greatest amount of aliment is found in the grain which has not been allowed to over-ripen in the field. Oat husk is very inferior to the bran of wheat. Toppings are seldom worth the price at which they are sold.

Indian Corn has been highly extolled as a fattening food for stock, and its chemical composition would seem to justify the high opinion which practical men have formed of its relative nutritive value. In the United States, the feeding of horses on Indian corn and hay has been found very successful; but in these countries oats will be found a more economical food. For fattening purposes Indian corn appears exceedingly well adapted, as it contains more ready-formed fat—4·5 per cent.—than is found in most of the other grains, and, on an average, 70 per cent. of starch. Pigs thrive well on this grain. The Galatz round yellow grain is somewhat superior to the American flat yellow seed.

Rye is not extensively cultivated in this country, but on the Continent it is raised in large quantities. In the north of Europe it forms a considerable proportion of the food of both man and the domesticated animals. In Holland it is commonly consumed by horses, but in England there has always been a prejudice against the use of this grain as food for the equine tribe. It has been highly recommended for dairy stock, five pounds of rye-meal, with a sufficiency of cut straw, constituting, it is stated, a dietary on which cows yield a maximum supply of milk. Irish-grown rye contains less starch, and more flesh-formers and oil, than the Black Sea grain.

Rice, although it forms the chief pabulum of nearly one-third of the human family, is the least nutritious of the common food grains. Rice-dust, an article obtained in cleaning rice for European consumption, is said to promote the flow of milk when given to cows. It is sold in large quantities in Liverpool, where, according to Voelcker, it often commands a higher price than it is worth.

Buckwheat is chiefly used as a food for game and poultry.

Malted Corn.—During a late session of Parliament a Bill was passed to exempt from duty malt intended to be used as food for cattle. As feeders may now become their own maltsters, it may be of some use to them to have here a *résumé* of this Bill :—

1. Any person giving security and taking out a licence may make malt in a malt-house approved by the Excise for the purpose; and all malt so made and mixed with linseed-cake or linseed-meal as directed, shall be free from duty.

2. The security required is a bond to Her Majesty, with sureties to the satisfaction of the Excise, not to take from any such malt-house any malt except duly mixed with material prescribed by the Act.

3. The malt-house must be properly named upon its door.

4. All malt made in it shall be deposited in a store-room, and shall be conveyed to and from the room upon such notice as the officer of Excise shall appoint.

5. The maltster shall provide secure rooms in his malt-house, to be approved in writing by the supervisor, for grinding the malt made by him in such malt-house, and mixing and storing the same when mixed; and all such rooms shall be properly secured and kept locked by the proper officer of Excise.

6. All malt before removal from the malt-house shall be ground and thoroughly mixed with one-tenth part at least of its weight of ground linseed-cake or linseed-meal, and ground to such a degree of fineness and in such manner as the commissioners shall approve, and mixed together in a quantity not less than forty bushels at a time in the presence of an officer of Excise.

7. The maltster shall keep account of the quantity of all malt mixed as aforesaid which he shall from time to time send out or deliver from his malt-house, with the dates and addresses of the person for whom such mixed malt shall be so sent or delivered.

8. If any person shall attempt to separate any malt from any material with which the same shall have been mixed as aforesaid, or shall use this malt for the brewing of beer or distilling of spirits, he shall forfeit the sum of £200.

9 and 10. The penalties of existing Acts are recited.

11. This Act shall continue and be in force for five years.

Some samples of malt and barley examined in May, 1865, by Dr. Voelcker for the Central Anti-Malt Tax Association, afforded the following results :—

	Barley marked No. 1.	Malt marked				
		No. 5.	No. 7.	No. 9.	No. 14.	No. 16.
Moisture	11·76	8·72	7·43	7·76	8·35	7·06
Sugar	3·75	4·29	5·48	7·85	9·46	9·86
Starch and dextrine	70·40	71·03	69·70	67·57	67·53	67·67
*Albuminous compounds (flesh-forming matters)	7·75	8·44	8·81	9·37	8·60	8·31
Woody fibre (cellular)	4·46	5·22	6·38	5·38	4·14	5·11
Mineral matter (ash)	1·88	2·30	2·20	2·07	1·92	1·99
	100·00	100·00	100·00	100·00	100·00	100·00
* Containing nitrogen	1·24	1·35	1·41	1·50	1·38	1·33

A great deal has been said and written in favor of malt as a feeding stuff, but I greatly doubt its alleged decided superiority over barley; and until the results of accurately conducted comparative experiments made with those articles incontestably prove that superiority, I think it is somewhat a waste of nutriment to convert barley into malt for feeding purposes. The gentlemen who verbally, or in writing, refer so favorably to malt, acknowledge, with one or two exceptions, that their experience of the article is limited. Mr. John Hudson, of Brandon, states that he made a comparative experiment, the results of which proved the superiority of malt. But, in fact, the only properly-conducted experiments to determine the relative values of malt and barley were those made some years ago by Dr. Thompson, of Glasgow, by the direction of the Government, and those recently performed by Mr. Lawes, both producing results unfavorable to the malt. The issue of Dr. Thompson's investigations proved that milch cows fed on barley yielded more milk and butter than when supplied with an equal weight of malt.

I do not deny the probability that malt, owing to its agreeable flavor and easy solubility, may be a somewhat better feeding stuff than barley; and that, weight for weight, it may produce a somewhat greater increase in the weight of the animals fed upon it: but although a pound-weight of malt may be better than a pound-weight of barley, I am quite satisfied that a pound's worth of barley will put up more flesh than a pound's worth of malt. Barley-seeds consist of water, starch, nitrogenous substances—such as gluten and albumen—fatty substances, and saline matter. The amount of starch is considerable, being sometimes about 70 per cent. In the process of malting (which is simply the germination of the seed under peculiar conditions), a portion of the starch is converted into sugar and gum, the grain increases in size and becomes friable when dried, and the internal structure of the seed is completely broken up. During these changes a partial decomposition of the solid matter of the seeds takes place, and a large amount of nutriment is dissipated, chiefly in the form of carbonic acid gas. From the results of the experience of the maltster, and of special experiments made by scientific men, it would appear that a ton of barley will produce only 16 cwt. of malt. Allowance must, however, be made for the difference between the amount of water contained in barley and in malt, the latter being much drier. According to Mr. E. Holden, the centesimal loss sustained in malting may be stated thus :—

Water	6·00
Organic matter	12·52
Saline matter	0·48
	100·00

Dr. Thompson* sets down the loss of nutriment (exclusive of that occasioned by kiln-drying), as follows :—

Carried off by the steep	1·5
Dissipated on the floor	3·0
Roots separated by cleaning	3·0
Waste...	0·5
	8·0

* Report to Government on feeding cattle with Malt, 1844.

P

We may say, then, that by the malting of barley we lose at least 2½ cwt. of solid nutriment out of every ton of the article, and this loss falls heaviest on the nitrogenous, or flesh-forming constituents of the grain. When there are added to this loss the expense of carting the grain to and from the malt-house, and the maltster's charge for operating upon it (I presume in this case that the feeder is not his own maltster), it will be found that two tons of malt will cost the farmer nearly as much as three tons of barley; and he will then have to solve the problem—*Whether or not malt is* 40 *or* 50 *per cent. more valuable as a feeding-stuff than barley.*

The difference in value between barley and malt is generally 14s. per barrel; but it is sometimes more or less, according to the supply and demand. Barley, well malted, will lose on the average 25 per cent. of its weight, the loss depending, to some extent, upon the degree to which the process is carried, and on the germinating properties of the barley. Barley malted for roasters ought not to lose more than 21 per cent. of its original weight—53 lbs. to the barrel. The heavier the barley the less it loses in malting; a barrel of 224 lbs., and value from 15s. to 16s., ought to produce a barrel of malt of 196 lbs., value 29s. to 30s.

If we deduct from the cost of a barrel of malt the amount of duty at present levyable upon it, the price of the article will be still nearly 50 per cent. greater than that of an equal weight of barley. The cheaper barley is the greater will be the relative cost of malt. The maltster's charge for converting a barrel of barley into malt is about 4s.; so that if the price of the grain be so low as 12s. per barrel, which it sometimes is, the cost of malting it would amount to 33 per cent. of its price. Then, the diminution in the weight of, and the cost of carting the grain, must be taken into account; and when the whole expense attendant upon the process of malting is ascertained, it will be found that I have not exaggerated in stating that a ton of malt costs as much as a ton and a half of barley.

If the consumer of malt germinate the seeds himself, he

may probably, if he require large quantities of the article, produce it at a somewhat cheaper rate than if he bought it from the maltster; but few persons who have the slightest knowledge of the vexatious restrictions of the Inland Revenue authorities would be likely to place his premises under the *espionage* of an excise officer.

As the superiority of malt over barley (if such be really the case) must be chiefly due to the looseness of its texture, which allows the juices of the stomach to act readily upon it, barley in a cooked state might be found quite as nutritious: it would not be fair to institute comparisons between dense hard barley-seeds and the easily soluble malted grains. During the cooking of barley a portion of the starch is changed into sugar, but in this case with only an inappreciable waste of nutriment. When the cooking process is continued for a few hours, a considerable amount of sugar is formed, and the barley acquires a very sweet flavor.

When the malt for cattle question was under discussion, I made a little experiment in relation to it, the results of which are perhaps of sufficient interest to mention:—Two pounds weight of barley-meal were moistened with warm water; after standing for three hours more water was added, and sufficient heat applied to cause the fluid to boil. After fifteen minutes' ebullition, a few ounces of the pasty-like mass which was produced were removed, thoroughly dried, and on being submitted to analysis yielded six per cent. of sugar. The addition of a small quantity of malt to barley undergoing the process of cooking will rapidly convert the starch into sugar.

Barley is naturally a well-flavored grain, and all kinds of stock eat it with avidity. It may be rendered still more agreeable if properly cooked, and this process will, by disintegrating its hard, fibrous structure, set free its stores of nutriment. I incline strongly to the opinion that barley, when well boiled, is almost, if not quite, as digestible as malt.

A serious disadvantage in the use of malt is, that it must be consumed, it is said, in combination with 10 per cent. of its weight of linseed-meal or cake. Now, malt is a very laxative food, and so is linseed; and if the diet of stock were largely made up of these articles the animals would, sooner or later, suffer from diarrhœa. In such case, then, the addition of bean-meal, or of some other binding food, would become necessary, and the compound of malt, linseed, and bean-meal thereby formed would certainly prove anything but an economical diet.

Malt Combs.—I should mention that a portion of the nutriment which the barley loses in malting passes into the radicles, or young roots, which project from the seeds, and are technically known by the term "combs," "combings," or "dust." At present these combs are separated from the malt, but if the latter be intended for feeding purposes this separation is unnecessary, and in such case the barley will not be so much deteriorated. The combs, which constitute about 4 per cent. of the weight of the malt, are sometimes employed as a feeding stuff. I have made an analysis of malt-combings for the County of Kildare Agricultural Society, and have obtained the following results:—

100 PARTS CONTAINED—

Water	8·42
*Flesh-forming (albuminous) substances	21·50
Digestible fat-forming substances (starch, sugar, gum, &c.)	53·47
Indigestible woody fibre	8·57
† Saline matter (ash)	8·04
	100·00
* Yielding nitrogen	3·44
† Containing potash	1·35
,, phosphoric acid	1.74

This article was sold as a manure at £3 6s. per ton—a sum for which it was not good value; but as a feeding substance it was probably worth £4 or £5 per ton. Its composition indicates a high nutritive power; but it is probable that its nitrogenous matters are partly in a low degree of elaboration, which greatly detracts from its alimental value.

In conclusion, then, I would urge the following points upon the attention of the farmer:—

1st. Before using malt for feeding purposes, wait until you learn the general results of the experience of other farmers with that article. The manufacture of malt for feeding purposes is rapidly on the decline, instead of, as had been anticipated, on the increase.

2nd. Should you experiment with barley and malt, use equal money's worth of each, and employ the barley in a cooked state.

3rd. Use malt-combings as a feeding stuff, and not as a manure. They are good value for at least £3 10s. per ton.

4th. Bear in mind that a ton of barley contains more saline matter than an equal weight of malt; consequently, that stock fed upon barley will produce a manure richer in potash and phosphates than those supplied with malt.

Leguminous Seeds.—The seeds of the bean, of the pea, and of several other leguminous plants, are largely made use of as food for both man and the domesticated animals. They all closely resemble each other in composition, but in that respect differ considerably from the grains of the *Cerealia*, for whilst the latter contain on an average 12 per cent. of flesh-formers, beans and peas contain 24 per cent. The flesh-forming constituent of the leguminous seeds is not gluten, as in the grain of the cereals, but a substance termed *legumin*, which so closely resembles the cheesy matter of milk that it has also received the name of *vegetable casein*. Indeed, the Chinese make a factitious cheese out of peas, which it is difficult to discriminate from the article of animal origin.

Beans are used as fattening food for cattle, for which purpose they should be ground into meal, as otherwise a large proportion of their substance would pass through the animal's body unchanged. It is not good economy to give a fattening bullock more than 3 or 4 lbs. weight per diem; a larger proportion is apt to induce constipation. The very small proportion of ready-formed fat, the moderate amount of starch, and the exceedingly high per-centage of flesh-formers which beans contain, prove that they are better adapted as food for beasts of burthen than for the fattening of stock. Oats, Indian corn, or oil-cake, will be found to produce a greater increase of meat than equal money's worth of beans or peas, and I would therefore recommend the restriction of leguminous seeds, under ordinary circumstances, to horses and bulls. It has been stated, on good authority, that when oats are given whole to horses, a large proportion passes unchanged through the animal's body, but that on the addition of beans, the oats are thoroughly digested.

COMPOSITION OF LEGUMINOUS SEEDS.

	Common Beans.	Foreign Beans.	Peas.	Lentils.	Winter Tares (foreign).
Water	13·0	14·5	14·0	13·0	15·5
Flesh-formers	25·5	23·0	23·5	24·0	26·5
Fat-formers	48·5	48·7	50·0	50·5	47·5
Woody fibre	10·0	10·0	10·0	10·0	9·0
Mineral matter	3·0	3·8	2·5	2·5	1·5
	100·00	100·00	100·00	100·0	100·0

Oil Seeds.—The seeds of a great variety of plants, such as the flax, hemp, rape, mustard, cotton, and sunflower, are exceedingly rich in oil, some of them containing nearly half their weight of that substance. Of these oil-seeds there are many which might with advantage be employed as fattening food, although one only—linseed—has come into general use for that purpose.

Rape-seeds closely resemble linseeds in composition, but they are considerably cheaper. They contain an acrid substance, but the large proportion of oil with which it is associated almost completely disguises its unpleasant flavor.

Linseed is one of the most valuable kinds of food which could be given to fattening animals. Its exceedingly high proportion of ready-formed fatty matter, the great comparative solubility of its constituents, and its mild and agreeable flavor, constitute it an article superior to linseed cake. The laxative properties of linseed are very decided; it should therefore be given only in moderate quantities. As peas and beans exercise, as I have already stated, a relaxing influence upon the bowels, a mixture of linseed and peas or beans would be an excellent compound, the laxative influence of the one being corrected by the binding tendency of the other. Linseed being one of the most concentrated feeding stuffs in use, it will be found an excellent addition to bulky food, such as chaff and turnips. Linseed oil has been used as a fattening food, but there is nothing to be gained by expressing seeds for the purpose of using their oil as a feeding material. When hay is scarce, and straw abundant, the latter may be made almost as nutritious as the former by mixing it with linseed, and steaming the compound. A stone of linseed and two cwt. of oat-straw chaff, when properly cooked, constitute a most economical and nutritious food.

Mr. Horne, who experimented with linseed two or three years ago, obtained results highly favorable to the nutritive value of that article. Six bullocks were selected, and each animal placed in a separate box. They were fed with cut roots—at first Swedes, then mangels and Swedes, and lastly, mangels alone: in addition, there were supplied to each 6 lbs. rough meadow-hay reduced to chaff, and 5 lbs. oil-cake, or value to that amount. They were divided into three lots, two in each. Lot 1 had 5 lbs. oil-cake for each animal; lot 2, barley and wheat-meal, equal in value to the 5 lbs. oil-cake; and lot 3, an equal money's worth of bruised linseed. The oil-cake cost £10 16s. per ton, the mixture of barley and wheat £8 15s. per ton, and the bruised linseed £13 per ton. The experiment lasted 112 days, and at its close the results, which proved very favorable to the bruised linseed, were as follows:—

	Increase in live weight.
Lot 1. Oil-cake	637 lbs.
Lot 2. Wheat and barley-meal...	667 lbs.
Lot 3. Bruised linseed	718 lbs.

During the 112 days each bullock consumed 5 cwt. oil-cake (or an equivalent amount of linseed or wheat and barley), 6 cwt. hay, and 90 cwt. of roots. The average increase in each animal's weight was 337 lbs. =224 lbs. *dead* weight. The economic features of this experiment are best shown in the following figures:—

FOOD CONSUMED.

	£	s.	d.
5 cwt. oil-cake, at 10s. 6d. per cwt.	2	12	6
6 cwt. hay, at 3s. per cwt.	0	18	0
16 weeks' attendance, at 6d. per week	0	8	0
	£3	18	6
Gained 16 stones per week, at 8s. per stone	6	8	0
Balance to pay for 90 cwt. of roots	2	9	6

The manure obtained afforded a good profit.

The seed-pods, or, as they are termed, the *bolls* of the flax, have been recommended as an excellent feeding stuff. They are not so nutritious as linseed, but they are cheaper, and when produced on the farm must be an economical food. Mr. Charley, an intelligent stock-feeder in the county of Antrim, and an eminent authority in every subject in relation to flax, strongly recommends the use of flax-bolls. He says:—

> The cost of rippling is considerable; but I believe, for every £1 expended, on an average a return is realised of £2, particularly on a farm-stead where many horses and cattle are regularly kept. The flax-bolls contain much more nourishment than the linseed-cake from which the oil has, of course, been expressed, and they form a most valuable addition to the warm food prepared during winter for the animals just named. I believe they have also a highly beneficial effect in warding off internal disease, owing, no doubt, to the soothing and slightly purgative properties of the oil contained in the seed. The change made in the appearance of the animals receiving some of the bolls in their steamed food is very apparent after a few weeks' trial; and the smoothness and sleekness of their shining coats plainly show the benefit derived. Is it not surprising, with this fact before our eyes, that many agriculturists—indeed, I fear the majority—persist in the old-fashioned system of taking the flax to a watering-place with its valuable freight of seed unremoved, and plunge the sheaves

under water, losing thereby, *in the most wanton manner*, rich feeding materials, worth from £1 to £3 per statute acre?

In the following table, the composition of all the more important oil-seeds is given:—

COMPOSITION OF OIL-SEEDS, ACCORDING TO DR. ANDERSON.

	Linseed.	Rape-seed.	Hemp-seed.	Cotton-seed (decorticated).
Water	7·50	7·13	6·47	6·57
Oil	34·00	36·81	31·84	31·24
Albuminous compounds (Flesh-formers)	24·44	21·50	22·60	31·86
Gum, mucilage, sugar, &c.	} 30·73	18·73	} 32·72	14·12
Woody-fibre		6·86		7·30
Mineral matter (ash)	3·33	8·97	6·37	8·91
	100·00	100·00	100·00	100·00

Fenugreek-seed is used very extensively in the preparation of "Condimental food." It is often given to horses out of condition. Sheep have been liberally supplied with this food, which, however, it is stated, communicates a disagreeable flavor to the mutton. It contains, according to Voelcker, the following:—

Water	11·994
Flesh-formers	26·665
Starch, gum, and pectin	37·111
Sugar	2·220
Fatty and oily matters	8·320
Woody fibre	10·820
Inorganic matter	2·870
	100·000

SECTION VII.

OIL-CAKES, AND OTHER ARTIFICIAL FOODS.

OIL-SEEDS, on being subjected to considerable pressure, part with a large proportion of their oil, the remaining part of that fluid, together with the various other ingredients of the seeds, constitute the substances so well known to agriculturists under the name of oil-cakes. These cakes contain a larger proportion of ready-formed fatty matter than is found in any other feeding stuff, and an amount of flesh-forming principles far greater than that yielded by corn, or even by beans; the manure, too, which is produced by the cattle fed upon some of them, is often good value for nearly half the sum expended on the food.

The principal kinds of oil-cake employed for feeding purposes are the following:—Linseed-cake, Rape-cake, and cotton-seed cake. Poppy cake is not much in use. Their average composition, deduced from the results of numerous analyses made by Voelcker, Anderson, and myself, are shown in the following table:—

AVERAGE COMPOSITION OF OIL-CAKES.

	Linseed Cake, English.	Rape Cake.	Decorticated Cottonseed Cake.	Poppy Cake.
Water	12	11	9	12
Flesh-forming principles	28	30	38	32
Oil	10	11	13	6
Gum, mucilage, &c.	34	30	23	30
Woody fibre	10	10	9	9
Mineral matter (ash)	6	8	8	11
	100	100	100	100

Linseed Cake.—Within the last quarter of a century great attention has been given to the feeding of stock, and the effects are observable in the improved quality and greatly increased weight of the animals. In the year 1839 the average weight of the horned beasts from Ireland sold in the London market was only 650 lbs., whereas at the present time their average weight is about 740 lbs. This remarkable advance in the production of meat is in great part due to the cattle being more liberally supplied with food, and that, too, of a more concentrated nature. The practice of feeding animals destined for the shambles exclusively on roots containing 90 and even 95 per cent. of water, which once prevailed so generally in this country, is now limited to the farmsteads of a few old-fashioned feeders; and the necessity for the admixture of highly-nutritious aliment with the bulky substances which form the staple food of stock is almost universally recognised.

Of concentrated foods used for fattening stock, none stands higher in the estimation of the farmer than linseed-cake, although it appears to me that the price of the article is somewhat too high in relation to its amount of nutriment, and that corn, if its price be moderate, is a more economical food. Straw, turnips, and mangels form the bone and sinew of the animals, and enable them to carry on the vital operations which are essential to their existence. Oil-cake and similar foods are supplemental, and contribute directly to the animal's increase, so that their nutritive value appears to be greater than it really is. If an animal were fed exclusively upon oil-cake, the greater part of it would be appropriated to the reparation of the waste of the body, and the rest would be converted into permanent flesh—the animal's "increase." The addition of straw would produce a still further increase in the animal's weight—an increase which would be directly proportionate to the amount of straw consumed. Thus it will be seen that, whatever the staple food may be, it will have to sustain the life of the animal, and will be principally expended for that purpose, whereas the supplemental food will be chiefly, if not entirely, made use of in increasing the weight of flesh. To me it appears manifestly incorrect to consider, as feeders practically do, the value of linseed-cake to be seven or eight times greater than that of oat-straw, and twenty times greater than that of roots. Let us assume the case of an animal fed upon roots, straw, and oil-cake. Seventy-five per cent. of its food, say, is expended in repairing the waste of its body, and 25 per cent. is stored up in its increase. Now, if the three kinds of food contributed proportionately to the reparation of the body and to its increase, the roots and straw would be found to possess a far higher nutritive value, in relation to the oil-cake, than is usually ascribed to them.

But it may be asked why straw, if it be relatively a much more economical feeding stuff than oil-cake, is not employed to the complete exclusion of the latter. I have already given an answer to such a question, namely, that animals thrive better on a diet composed partly of bulky, partly of concentrated aliments. This much, however, is certain, that animals can be profitably fed upon roots and straw, whilst it is equally certain that to feed them upon oil-cake alone (assuming them to thrive upon such a diet) would entail a very heavy loss upon the feeder. At the same time it must be admitted that the oil of the linseed-cake exercises in all probability a beneficial influence on the digestion of the animal, so that the nutritive value of the article may be somewhat higher than its mere composition would indicate.

The quantity of oil-cake given to fattening stock varies from 2 lbs. to 14 lbs. per diem. I believe there is no greater mistake made by feeders than that of giving excessive quantities of this substance to stock. If their object in so doing be to enrich their manure-heap, they would find it far more economical to add the cake directly to the manure—or rather of adding rape-cake to it, for this variety of cake is fully as valuable for manurial purposes as the linseed-cake, and is nearly 50 per cent. cheaper. A larger quantity of oil-cake than 7 lbs. daily should not be given to even the largest-sized milch cows or fattening bullocks. If a larger amount be employed, it will pass unchanged through the animal's body. Young cattle may with advantage be supplied with from 1 to 3 lbs., according to their size, and from $\frac{1}{2}$ to 1 lb. will be a sufficient quantity for sheep. Intelligent feeders have remarked, that cattle which had been always supplied with a moderate allowance of this food fattened more readily upon it, during their finishing stage, than did stock which had not been accustomed to its use.

Adulteration of Linseed Cake.—The great drawback to the use of linseed-cake is the liability of the article to be adulterated. The sophistication is sometimes of a harmless nature, if we except its injurious effect on the farmer's pocket; but not unfrequently the substances added to the cakes possess properties which completely unfit them to be used as food. Amongst the injurious substances found in linseed and linseed-cake I may mention the seeds of the purging-flax, darnel, spurry, corn-cockle, curcus-beans, and castor-oil beans. Several of these seeds are highly drastic purgatives, and they have been known to cause intense inflammation of the bowels of animals fed upon oil-cake, of which they composed but a small proportion. Amongst the adulterations of linseed-cake, which lower its nutritive value without imparting to it any injurious properties, are the seeds of the cereals and the grasses, bran, and flax-straw. Little black seeds belonging to various species of *Polygonum*, are very often present in even good cakes; they are very indigestible, but otherwise are not injurious. Rape-cake is stated to be occasionally used as adulterant of the more costly linseed, but I have never met with an admixture of the two articles.

The only way in which a correct estimate of the value of linseed-cake can be arrived at is by a combined microscopical and chemical analysis; but as the feeder is not always disposed to incur the cost of this process, he should make himself acquainted with the characteristic of the genuine cake, in order to be able to discriminate, as far as possible, between it and the sophisticated article. I will indicate a few of the more prominent features of cake of excellent quality, and point out a few simple and easily-performed tests, which may serve to detect the existence of gross adulteration. Good cake is hard, of a reddish-brown color, uniform in appearance, and possesses a rather pleasant flavor and odour. The adulterated cake is commonly of a greyish hue, and has a disagreeable odour. A weighed quantity of the cake—say 100 grains—in the state of powder should be formed into a paste with an ounce of water;

if it be good, the paste will be light colored, moderately stiff, and endowed with a pleasant odour and flavor. If the paste be thin, the presence of bran, or of grass seeds, is probable. The latter are easily seen through a magnifying-glass; indeed, most of them are readily recognisable by the unassisted eye: they may, therefore, be picked out, and their weight determined. Sand—a frequent adulterant—may be detected by mixing a small weighed quantity of the powdered cake with about twelve times its weight of water, allowing the mixture to stand for half an hour, and collecting and weighing the sand which will be found at the bottom of the vessel employed. If there be bran present it will be found lying on the sand, and its structure is sufficiently distinct to admit of its detection by a mere glance. There are a great variety of linseed-cakes in the market, of which the home-made article is the best. On the Continent the oil-seeds are subjected to the action of heat in order to obtain from them a greater yield of oil. Their cakes, therefore, contain less oil, and their flesh-forming principles are less soluble, in comparison with British linseed-cake. Next to ur home-made oil-cakes, the American is the best. Indeed, I have met with some American cakes which were equal to the best English.

Rape Cake.—The use of rape-cake was limited almost completely to the fertilising of the soil until the late Mr. Pusey, in a paper published in the tenth volume of the *Journal of the Royal Agricultural Society of England*, advocated its employment as a substitute for the more costly linseed-cake. The recommendation of this distinguished agriculturist has not been disregarded; and since his time the use of this cake as a feeding stuff has been steadily on the increase, and at the present time its annual consumption is not far short of 50,000 tons.

In relation to the nutritive value of rape-cake there exists considerable diversity of opinion. Certain feeders assert that animals fed upon it go out of condition; others, whilst admitting that stock thrive upon it, maintain the economic superiority of linseed-cake; whilst a third set believe rape-cake to be the most economical of feeding-stuffs. How are we to account for these great differences of opinion—not amongst *theorists*, be it observed, but amongst practical men? It is not difficult to explain them away satisfactorily. Rape-cake and linseed-cake are about equally rich in muscle and fat-forming principles; and, supposing both to be equally well-flavored, there can be no doubt but that one is just as nourishing as the other. But it so happens that a large proportion of the rape-cake which comes into the British market possesses a flavor which renders it very disagreeable to animals. One variety—namely, the East Indian—is almost poisonous, whilst the very best kind is slightly inferior to linseed-cake. Now, if an experiment with a very inferior kind of rape-cake and a good variety of linseed-cake were tried, who can doubt but that the results would be very unfavorable to the former article? Mr. Callan,* of Rathfarnham, county Dublin; Mr. Bird,† of Renton Barns, and some other feeders, who found rape-cake to be worse than useless, experimented, in all probability, with an adulterated article, for they do not appear to have had the cake analysed. On the other hand, those whose experience with rape-cake has proved favorable, must have employed the article in a genuine state, fresh, and moderately well-flavored. It is noteworthy that amongst the advocates for the use of rape-cake as a substitute—partly or entirely—for the more costly linseed-cake, are to be found the most successful feeders in England and Scotland. Horsfall, Mechi, Lawrence, Bond, Hope, and many other feeders of equal celebrity, have assigned to rape-cake the highest place, in an economic point of view, amongst the concentrated feeding stuffs. Mr. Mechi says :—" I invariably give to all my animals as much rape-cake as they choose to eat, however abundant their roots or green food may be. It pays in many ways, and not to do this is a great pecuniary mistake. Even when fed on green rape, they will eat rape-

* *Monthly Agricultural Review*, Dublin, February, 1859.
† *Transactions of the Highland and Agricultural Society of Scotland*, October, 1858.

cake abundantly. My cattle are now under cover, eating the steamed chaff, rape-cake, malt-combs, and bran, all mixed together in strict accordance with the proportions named by Mr. Horsfall in the *Journal of the Royal Agricultural Society*, vol. xviii., p. 150,* which I find by far the most profitable mode of feeding bullocks and cows." Mr. Hope, of Edinburgh, states that rape-cake is the best substitute for turnips, and that, excepting cases where spurious kinds had been used, he never knew bullocks or milch cows to refuse it. This gentleman states that it is best given in combination with locust-beans, or a mixture of locust-beans and Indian corn; and suggests the proportions set down in the tables as the best adapted for lean cattle; but I think about two-thirds of the quantities would be quite sufficient.

	Feed per week. lbs.	Per week. s. d.
Rape-cake at £5 15s. per ton	8	2 10½
Do. do.	10	3 7
Mixture of two-thirds rape-cake and one-third locust-beans £6	8	3 0
Do. do.	10	3 9
Rape-cake, locust-beans, and Indian corn in equal proportions	8	3 2⅓
Do. do.	10	3 11¼

An intelligent Scotch dairy farmer bears the following testimony in favor of this cake:—

I have tried pease-meal, bean-meal, oat-meal, and linseed-cake, and after carefully noting the results, I consider rape-cake, weight for weight, at least equal to any of them for milch cows; and if I give the same money value for each, I get at least one-third more produce, and the butter is always of a very superior quality. Two years ago, I took some of my best oats (41 lbs. per bushel), and ground them for the cows, and although I was at about one-third more expense, I lost fully one-third of the produce that I had by using rape-cake. I always dissolve it by pouring boiling water on it, and give each cow 6 lbs. daily. I have tried a larger quantity, and found I was fully repaid for the extra expense. I generally use it the

* 3 lbs. of rape-cake, ¾ lb. malt combs, ¾ lb. bran, steamed together with a sufficient quantity of straw.

most of the summer, but always during the spring months. A number of my neighbours who have tried it all agree that it is the best and cheapest feed for milch cows they have used.—*North British Agriculturist*, Edinburgh, February 29, 1860.

The best kinds of rape-cake come from Germany and Denmark. When neither too old nor too fresh, and of a pale-green color, these foreign cakes are tolerably well-flavored, and are but slightly inferior to good linseed-cake. Most varieties of this cake, however, contain a small proportion of acrid matter, which often renders them more or less distasteful to stock, more particularly to cattle. This substance may be rendered quite innocuous by steaming or boiling the cake; either of these processes will also, according to Mr. Lawrence, destroy the disagreeable flavor which mustard-seed—a frequent adulterant of rape-cake—confers upon that article. Molasses or treacle is an excellent adjunct to the cake, as it serves in a great measure to correct its somewhat unpleasant flavor. Carob, or locust-beans, answer, perhaps better, the same purpose. It is better, as a general rule, to give less rape-cake than linseed-cake, unless the pale-green kind to which I have referred is obtainable; that variety may be largely employed. The animals should be gradually accustomed to its use. At first, in the case of bullocks, they should get only 1 lb. per diem, and the quantity should be gradually increased to about 4 lbs.; but I would not advise, under any circumstances, a larger daily allowance than 5 lbs. Given in moderate amounts, it will, supposing it to be of fair quality, be found to give a better return in meat than almost any other kind of concentrated food; and, what is of great importance, it will not injuriously affect the animal's health. "Our experience of the use of rape-cake," says Mr. Lawrence, "thus used (cooked), extends over a period of ten years of feeding from 20 to 24 bullocks annually. We have not had a single death during that period, and the animals have been remarkably free from any kind of ailment."

Rape-cake of good quality possesses a dark-green color

(the greener the better), and when broken exhibits a mottled aspect—yellowish and dark-brown spots. Sometimes a tolerably good specimen has a brownish color; but the German and Danish cakes are always of a greenish hue. The odor is stronger than that of linseed-cake, and differs but little from that of rape-oil. The only serious adulteration of rape-cake is the addition to it of mustard-seed—sometimes accidentally— less frequently, as I believe, intentionally. This sophistication admits of easy detection. Scrape into small particles about half an ounce of the cake, add six times its weight of water, form the solid and liquid into a paste, and allow the mixture to stand for a few hours. If the cake contain mustard the characteristic odor of that substance will be evolved, and its intensity will afford a rough indication of the amount of the adulterant. As some specimens of genuine rape-cake possess a somewhat pungent odor, care must be taken not to confound it with that of mustard; but, indeed, it is not difficult to discriminate the latter. The paste of rape-cake which contains an injurious proportion of mustard, has a very pungent flavor. Rape-cake improves somewhat if kept for say six months; but old cake is worse than the fresh article.

Cotton-seed Cake is one of the most valuable feeding stuffs that have come into use of late years. Its chemical composition shows it to be about equal to that of the best linseed-cake, and as its price is much lower than that of the latter, it may be fairly considered a more economical food. These remarks apply only to the shelled, or decorticated seed-cake, for the article prepared from the whole seed is of very inferior composition, and should never be employed. The use of the cake made from the whole seed has proved fatal in many instances, not from its possessing any poisonous quality, but in consequence of its hard, indigestible husk, accumulating in, and inflaming, the animal's bowels.

The composition of this cake varies somewhat. The following analysis of a sample from one of the Western States of

ARTIFICIAL FOODS. 243

North America, imported by Messrs. G. Seagrave and Co., of Liverpool, was made by me :—

COMPOSITION OF DECORTICATED COTTON-SEED CAKE.

Water	8·20
Oil	10·16
Albuminous, or flesh-forming principles	40·25
Gum, sugar, &c.	21·10
Fibre	9·23
Ash (mineral matter)	11·06
	100·00

In some specimens so much as 16 per cent. of oil has been found. The purchaser of cotton-seed cake should be certain that it is not old and mouldy, which is frequently the case. The recently prepared cake has a very yellow color, which becomes fainter as the cake becomes older. Freshness is a very desirable quality in nearly every kind of cake. I have known animals to have a greater relish for, and thrive better upon, home-made linseed-cake than upon cake of foreign manufacture of superior composition, but of greater age.

Palm-nut Meal, or *Cake* is a very valuable fattening food. It is extremely rich in ready-formed fatty matters, but at the same time it is not very deficient in albuminous substances. Its strong flavor is rather a drawback to its use in the case of all the farm animals, except pigs. This difficulty may, however, be got over by using the cake in moderate quantities, and by combining it with other food possessed of a good flavor. Reports of practical trials made with this food appear to have almost uniformly given very favorable results. This food is only three or four years in use. The first samples that came into my hand were richer in fatty matters than those which I have recently examined. The average results of eight analyses made from 1864 to 1866 were as follows :—

100 PARTS CONTAINED—

Water	7·48
Albuminous matters	17·26
Fatty substances	21·59
Gum, sugar, &c.	32·14
Fibre	17·18
Mineral matter	4·35
	100·00

This year I have not found more than 17 per cent. of fat in any sample of palm-nut cake. One specimen which I analysed for Mr. J. G. Alexander, seed merchant, of Dublin, had the following composition :—

Water	9·24
Albuminous matters	19·28
Fatty matters	9·36
Gum, starch, fibre, &c.	53·22
Mineral matters	8·90
	100·00

But although inferior samples are occasionally met with, I may say of palm-nut cake that on the whole it is a food which deserves to be largely used, and which at its present price is the most economical source of fat. To milch-cows and fattening cattle about 3 lbs. per diem may be given; ¼ lb. will be sufficient for young sheep, whilst pigs may be very liberally supplied with this food.

The *Locust, or Carob Bean*, is now largely used by the stock-feeder. It is extremely rich in sugar, and is therefore an excellent fattening and milk-producing food. It is used largely in the preparation of the sweet kinds of artificial food for cattle. It is not well adapted for young animals, owing to its deficiency of albuminous matters. The following analysis shows the average composition of this food :—

ARTIFICIAL FOODS. 245

Water	14
Sugar	50
Albuminous matters	8
Oil	1
Gum, &c.	20
Woody fibre	5
Ash	2
	100

Dates have been used, but only in very small quantities, as cattle food. Their composition is not constant, some samples being greatly inferior in nutritive power to others; they are rich in sugar, and if they were obtained in sufficient quantities they might, like carob-beans, come into general use with the stock-feeder. They contain about 2 per cent. of flesh-formers. 10 per cent. of fat-formers (chiefly sugar), and 2 per cent. of mineral matter.

Distillery and brewery dregs (or wash) are chiefly used by dairymen. According to Dr. Anderson, an imperial gallon (700,000 grains) of distillery wash (from a distillery near Edinburgh) contained 4,130 grains of organic matter, and 276 grains of mineral substances. He considers that 15 gallons of this stuff were equal in nutritive materials to 100 pounds of turnips. The following is the centesimal composition of brewery wash:—

Water	75·85
Albuminous matters	0·62
Gummy matters	1·06
Other organic matter (husks, &c.)	21·28
Mineral matters	1·19
	100·00

Molasses constitute a very fattening food, sometimes, but not often, given to stock. Treacle and molasses are composed of non-crystallisable sugar, cane-sugar, water, and saline and other impurities. The composition of average specimens of molasses, as imported, is as follows:—

Cane-sugar	50
Non-crystallisable sugar and grape-sugar	25
Water, saline matter, and organic impurities	25
	100

If admitted duty free, molasses would be a much more economical food than it now is, but at its present price it must be regarded as a mere flavoring food.

Mr. T. Cooke Burroughs, a West Suffolk feeder, who used treacle in 1864, gives the following mode of mixing it with other food :—

> My plan has been (and is still carried on) to give to each bullock per day (divided into three meals) one pint of treacle dissolved in two gallons of water, and sprinkled, by means of a garden water-pot, over four bushels of cut chaff (two-thirds straw and one-third hay) amongst which a quarter of a peck of meal (barley and wheat) is mixed, the animals also having free access to water. The cost of the treacle and meal together is about 3s. per bullock per week. My bullocks (two-year old Shorthorns) have grown and thrived upon the above diet to my utmost satisfaction; and even during the present dry and warm weather they evince no lingering after roots or grass. I am well aware that the use of treacle for neat stock is no new discovery of my own, as I learnt the system while on a visit to a friend in Norfolk, where some graziers have used it in combination with roots during many years past. Perhaps flax-seed (linseed) boiled into a jelly and used in a similar way, may be a more profitable "substitute for roots" than treacle; but the preparation of it is attended with more expense and trouble.

SECTION VIII.

CONDIMENTAL FOOD.

ALTHOUGH every farmer may not have used, there are few who have not heard of "Thorley's Condimental Food for Cattle." This nostrum is a compound of some of the ordinary foods with certain well-known aromatic and carminative substances. It possesses a very agreeable flavor, and it is therefore much relished by horses, and indeed by every kind of stock. The price of this compound was at first so much as £60 per ton; but owing to competition, and perhaps to the attacks made upon the enormously high price of this article, it is now to be obtained at prices varying from £12 to £24 per ton.

The inventor of condimental food, and the numerous fabricators of that compound, claim for it merits of no ordinary nature. Its use, they assert, not only maintains the animals fed upon it in excellent health, but it also exercises so remarkable an action upon the adipose tissues that fat accumulates to an immense extent. Moreover, it is said that an animal supplied with a very moderate daily modicum of this wonderful compound, will consume less of its ordinary food, though rapidly becoming fat.

Now, if these assertions were perfectly, or even approximatively, true, Mr. Thorley would be well deserving of a niche in the temple of fame, and stock-feeders would ever regard him as a benefactor to his own and the bovine species; but I fear that Mr. Thorley's imagination outstripped his reason when he described in such glowing terms the wonderful virtues of his tonic food.

Mr. J. B. Lawes, of Rothamstead, than whom there is no more accurate experimenter in agricultural practice, states that he made many careful trials with Thorley's food, and that he never found it to exercise the slightest influence upon the nutrition of the animals fed upon it. In his report upon this subject, Mr. Lawes, after describing the experiments which he made, sums up as follows:—

> There is nothing therefore in the above results to recommend the use of Thorley's condiment with inferior fattening food, to those who feed pigs for profit. In fact, the following balance-sheet of the experiment shows that, in fattening for twelve weeks, there was a balance of £1 10s. 11d. in favor of the lot fed without Thorley's food, notwithstanding that one of the pigs in that lot did badly throughout the experiment, as above stated.

LOT 1.—WITH BARLEY-MEAL AND BRAN.

	£	s.	d.
4 pigs bought in at 41s. 6d. each	8	6	0
1,860¾ lbs. barley, at 37s. 6d. per quarter of 416 lbs., including grinding...	8	7	8¾
1,024¾ lbs. bran at 5s. 6d. per cwt.	2	10	3¾
	19	4	0½
88 stone 5 lbs. of pork sold at 4s. 4d. per stone, sinking the offal	19	4	0¼

LOT 2.—WITH BARLEY-MEAL, BRAN, AND THORLEY'S FOOD.

	£	s.	d.
4 pigs bought in at 41s. 6d. each	8	6	0
1,862¾ lbs. barley, at 37s. 6d. per quarter of 416 lbs., including grinding	8	7	10¾
1,020¾ lbs. bran at 5s. 6d. per cwt.	2	10	1½
105 lbs. Thorley's food at 40s. per cwt.	1	17	6
	21	1	5¾
90 stone 1 lb pork sold at 4s 4d. per stone, sinking the offal	19	10	6½
	1	10	11¼

The results of these experiments with pigs, in which Thorley's condiment was used with inferior fattening food, may be summed up as follows:—

1. The addition of Thorley's condimental food increased the amount of food consumed by a given weight of animal within a given time.

2. When Thorley's condiment was given it required more food to produce a given amount of increase in live-weight.

3. In fattening for twelve weeks there was a difference of £1 10s. 11d. on the lot of 4 pigs in favor of barley-meal and bran alone, over barley-meal, bran, and Thorley's food in addition.

At a meeting of the Council of the Royal Agricultural Society of England, held some time ago, the subject of the nutrimental value of condimental cattle food was discussed. As there is scarcely any kind of quackery, from spirit manifestations to Holloway's pills, that has not got its believers, there were, as might have been anticipated, some voices raised at this meeting in favor of Thorley's food; but the *sense* of the meeting was decidedly against it. Professor Simonds pronounced it to be worthless.

Although the greater number of equine proprietors and feeders of stock are too sensible to throw their money away in the purchase of those costly foods, still there are by no means an insignificant number who employ it, under the idea that it preserves the health of the animals; these stuffs are also highly appreciated by many grooms and herds. Now, for the information of all believers, I may state that there is no mystery whatever in the nature of condimental cattle foods. They consist in substance of such matters as linseed-cake, Indian corn, rice, bean-meal, locust-beans, and malt-combings. These substances are flavored by the addition of turmeric-root, ginger, coriander-seed, carraway-seed, fenugreek-seed, aniseed, liquorice, and similar substances. In addition to the nutritive and flavorous articles employed in the manufacture of these foods, purely medicinal substances are also made use of with the idea that they would prove useful in maintaining the health and stimulating the appetite of the animals. These medicinal ingredients constitute but a small proportion of the compound, although they add considerably to the cost of manufacture. The following is a formula for a condimental food, which in every respect will be found fully equal, if not superior, to the ordinary high-priced articles.

	cwt.	qrs.	lbs.
Linseed-meal, or cake	7	0	0
Locust beans (ground)	8	0	0
Indian corn	4	1	0
Powdered turmeric	0	1	4
Ginger	0	0	3
Fenugreek-seed	0	0	2
Gentian	0	0	10
Cream of tartar	0	0	2
Sulphur	0	0	20
Common salt	0	0	10
Coriander-seed	0	0	5

One ton.

A ton of condimental food manufactured according to this formula will cost only about the same amount as an equal weight of linseed, and will produce an effect fully equal to that of the food which at one time was sold at £60 per ton.

Whatever may be the medicinal virtues of these foods, or however appropriate the term "condimental" which has been applied to them, it is quite certain that their whilom designation "concentrated" was a misnomer. Their composition shows that they possess a degree of nutritive power considerably below that of linseed-cake, and but little, if at all, superior to that of Indian corn.

The following analytical statement, which I published some years ago, will give an insight into the nature of these articles:—

ANALYSES OF CONDIMENTAL FOOD.

	Thorley's.	Bradley's.
Water	12·00	12·09
Nitrogenous, or flesh-forming principles	14·92	10·36
Oil	6·08	5·80
Gum, sugar, mucilage, &c.	56·86	60·21
Woody fibre	5·46	5·32
Mineral matter (ash)	4·68	6·22
	100·00	100·00

As a ton of linseed-cake contains a greater amount of nutriment than an equal quantity of condimental food, the latter should be clearly proved to possess very valuable specific virtues, in order to induce the feeder to use it extensively. Cattle and horses out of condition may be benefited by its carminative and tonic properties; but if they are, it surely must be a bad practice to feed healthy animals upon a substance which is a remedy in disease. It is asserted, and probably with some degree of truth, that when dainty, over-fed stock loathe their food, they are induced to eat greedily by mixing the "condimental" with their ordinary food. If such really be the case, let the feeder compound the article himself, and effect thereby a saving of perhaps 50 or 80 per cent. in the cost of it. A good condimental food, rich in actual nutriment, and pleasantly flavored, is no doubt a compound which might be used with advantage; but it should be sold at a moderate and fair price.

SECTION IX.—ANALYSES OF THE ASHES OF PLANTS.

(Extracted from the Author's "Chemistry of Agriculture.")

Those numbers marked with an asterisk refer to 100 parts of the substance in its natural or undried state; the remaining numbers refer to 100 parts when dried.

	Rape Seed.	Flax. Stalk.	Flax. Seed.	Peas.	Kidney Beans.	White Turnip Seed.	Turnip Bulb (Swede).	Cucumber.	Mangel Wurtzel Seed.	Potatoes (tubers).	Hop Flowers.
Potash	25·18	34·96	32·55	43·09	36·83	21·91	39·82	47·52	16·08	35·15	19·41
Soda	...	...	2·51	...	18·40	1·23	10·86	...	6·86	5·77	0·70
Lime	12·91	15·87	9·45	4·77	7·75	17·40	12·75	6·31	13·42	2·14	14·15
Magnesia	11·39	3·68	16·23	8·06	6·33	8·74	4·68	4·26	15·22	2·69	5·34
Sesquioxide of Iron	0·62	4·84	0·38	...	2·24	1·95	0·89	...	0·40	1·79	2·41
,, Manganese	...	...	...	...	...	...	...	...	...	...	...
Sulphuric Acid	0·53	4·99	1·43	0·44	3·96	7·10	13·15	4·60	3·64	3·29	8·28
Muriatic Acid	0·11	...	...	1·96	...	...	3·68	...	...	...	2·26
Carbonic Acid	2·20	13·39	...	...	...	0·82	...	...	13·85	17·14	11·01
Phosphoric Acid	45·95	8·48	35·99	40·56	11·60	40·17	6·69	18·03	13·35	20·70	14·64
Silica	1·11	5·60	1·46	0·79	4·09	0·67	7·05	7·12	1·86	3·00	18·56
Chloride of Potassium	...	7·65	...	...	...	...	...	4·19	...	1·84	...
Chloride of Sodium	...	0·54	...	...	2·80	...	...	9·06	15·30	6·49	2·95
Total	100·00	100·00	100·00	99·67	100·00	99·99	99·57	100·09	99·98	100·00	99·71
Per-centage of Ash	4·51	5·00	3·05	5·21	0·68*	3·98	7·60	0·63*	6·58	6·05

VEGETABLE FOODS.

ANALYSES OF THE ASHES OF PLANTS.

The number marked with an asterisk refers to 100 parts of the substance in its natural or undried state; the remaining numbers refer to 100 parts when dried.

	Cauli-flowers.	Hopeton Oats (Grain).	Potato Oats (Grain).	Husks of Potato Oats.	Rye. Grain.	Rye. Straw.	Hay.	Grasses (in flower). Bromus erectus.	Grasses (in flower). Lolium perenne.	Grasses (in flower). Annual Rye-grass.	Grasses (in flower). Avena flavescens.
Potash	34·39	20·65	⎱31·56	2·23	31·76	17·36	20·80	20·33	24·67	28·99	36·06
Soda	14·79	...	⎰	8·97	4·45	0·31	10·85	...	...	0·87	0·73
Lime	2·96	10·28	5·32	4·30	2·92	9·06	8·24	10·38	9·64	6·82	7·98
Magnesia	2·38	7·82	8·69	2·35	10·13	2·41	4·01	4·99	2·85	2·59	3·07
Sesquioxide of Iron	1·69	3·85	0·88	0·32	0·82	1·36	1·83	0·26	0·21	0·28	2·40
,, Manganese	...	0·42	...	...	...	...	...	...	...	...	...
Sulphuric Acid	11·16	...	...	4·30	1·46	0·83	2·11	5·46	5·20	3·45	4·00
Muriatic Acid	...	...	...	...	...	0·46	...	...	...	...	...
Carbonic Acid	...	...	...	...	...	...	0·68	0·55	0·49	...	...
Phosphoric Acid	27·85	50·44	49·19	0·66	47·29	3·82	15·43	7·53	8·73	10·07	9·31
Silica	1·92	4·40	1·87	74·18	0·17	64·50	30·01	38·48	27·13	41·79	35·20
Chloride of Potassium	...	1·03	...	...	...	...	...	10·63	13·80	...	...
Chloride of Sodium	2·86	...	0·35	2·39	...	...	5·09	1·38	7·25	5·11	1·25
Total	100·00	98·89	97·86	99·70	100·00	100·11	99·05	99·99	99·97	99·97	100·00
Per-centage of Ash	0·71*	...	2·22	...	2·30	2·60	...	5·21	7·54	6·45	5·20

ANALYSES OF THE ASHES OF PLANTS.

Those numbers marked with an asterisk refer to 100 parts of the substance in its natural or undried state; the remaining numbers refer to 100 parts when dried.

	Broccoli.		Cow Cabbage.			Kohl-rabi, from chalk soil.		Wheat (Grain).	Wheat.		Barley.	
	Root.	Leaves.	Leaves.	Stalk.	Leaves.	Tuber.			Grain.	Straw.	Grain.	Straw.
Potash	47·16	22·10	40·86	40·93	9·31	36·27	29·51	25·92	10·78	32·02	14·37
Soda	...	7·55	2·43	4·05	...	2·84	10·61	...	...	1·21	0·28
Lime	4·70	28·44	15·01	10·61	30·31	10·20	0·99	3·80	2·44	3·39	8·50
Magnesia	3·93	3·43	2·39	3·85	3·62	2·36	10·60	12·27	3·23	10·99	1·70
Sesquioxide of Iron ...	...	...	0·77	0·41	5·50	0·38	...	1·12	0·54	0·15	0·20
,, Manganese	...	...	...	...	...	...	...	...	...	...	...
Sulphuric Acid ...	10·35	16·10	7·27	11·11	10·63	11·43	0·09	...	1·77	...	2·22
Muriatic Acid ...	...	...	...	...	...	...	...	...	...	...	...
Carbonic Acid ...	...	...	16·68	6·33	8·97	10·24	...	4·43	6·01	0·48	1·25
Phosphoric Acid ...	25·83	19·81	12·52	19·57	9·43	13·46	47·55	43·44	3·69	29·92	4·22
Silica	1·81	2·83	1·66	1·04	9·57	0·82	0·11	7·16	64·84	21·12	62·89
Chloride of Potassium...	6·22	...	...	...	5·99	...	...	1·03	3·96	...	...
Chloride of Sodium ...	a trace	...	...	2·08	6·66	11·90	0·54	.	0·42	0·72	4·37
Total	100·00	100·26	99·99	99·98	99·99	99·90	100·00	99·17	99·68	100·00	100·00
Per-centage of Ash ...	1·01*	1·70*	0·70*	1·24*	18·54	8·09	2·32	1·645	5·252	2·22	5·49

APPENDIX.

WHILST this Work was passing through the press, a valuable Report on Agricultural Statistics was issued by the Board of Trade. The following statistics, collected from this Report, are here given, because they modify the statements made in page 5:—

POPULATION, AREA, ACREAGE UNDER CROPS, ETC., AND NUMBER OF LIVE STOCK, IN THE UNITED KINGDOM IN 1867.

	England.	Wales.	Scotland.	Ireland.	Isle of Man.	Channel Islands.		Total for United Kingdom
						Jersey.	Guernsey, &c.	
Population (1866) ...	20,276,494	1,187,103	3,136,057	5,571,971	52,469	55,613	35,365	30,315,072
Area (in Statute Acres)	32,599,397	4,734,486	19,639,377	20,322,641	180,000	28,717	17,967	77,513,585
Under Corn Crops ...	7,399,347	521,404	1,364,029	2,115,137	27,039	2,827	2,157	11,431,940
,, Green Crops ...	2,691,734	138,387	668,042	1,432,252	12,670	5,636	3,075	4,951,796
,, Bare Fallow ...	753,210	86,257	83,091	26,191	1,990	2,550	709	953,998
,, Grass—Clover, &c., under Rotation	2,478,117	300,756	1,211,101	1,658,451	26,884	3,250	874	5,679,433
Permanent Pasture, not broken up in Rotation*	9,545,675	1,472,359	1,953,285	10,057,072	15,915	6,092	6,143	22,156,541
Per-centage of Acreage:†								
Under Corn Crops	32·3	20·7	31·1	13·6	32·0	13·9	16·7	25·1
,, Green Crops	11·7	5·3	15·3	9·2	15·0	27·6	23·7	10·9
,, Bare Fallow	3·3	3·4	1·9	·2	2·4	12·5	5·5	2·1
,, Grass—Clover, &c., under Rotation	10·8	11·9	27·7	10·7	31·8	16·0	6·7	12·4
Permanent Pasture‡	41·6	58·5	24·0	64·7	18·8	30·0	47·4	48·7
Number of Cattle	3,469,026	544,538	979,470	3,702,378	18,672	10,081	7,308	8,731,473
,, of Sheep	19,798,337	2,227,161	6,893,603	4,826,015	70,938	529	1,348	33,817,951
,, of Pigs	2,548,755	229,917	188,307	1,233,893	7,706	5,804	6,718	4,221,100
Number of Live Stock to every 100 Acres under Crops, Fallow, and Grass:—								
Cattle	15·1	21·6	22·4	23·8	22·1	49·5	56·4	19·2
Sheep	86·3	88·4	157·4	31·1	84·0	2·6	10·4	74·3
Pigs	11·1	9·1	4·3	7·9	9·1	28·5	51·8	9·3

* Exclusive of heath or mountain land.   † The per-centage of acreage is exclusive of Hops in Great Britain, and Flax in Ireland.
‡ Including under Flax, 251,105 acres.

www.ingramcontent.com/pod-product-compliance
Lightning Source LLC
Chambersburg PA
CBHW021346230426
43666CB00006B/425